MATRICES FOR ENGINEERS

MATRICES FOR ENGINEERS

Allan D. Kraus
Naval Postgraduate School
Monterey, California

⊙ HEMISPHERE PUBLISHING CORPORATION
A subsidiary of Harper & Row, Publishers, Inc.

Washington   New York   London

DISTRIBUTION OUTSIDE NORTH AMERICA
SPRINGER–VERLAG

Berlin   Heidelberg   New York   London   Paris   Tokyo

MATRICES FOR ENGINEERS

1 2 3 4 5 6 7 8 9 0   B R B R   8 9 8 7

This book was set in Times Roman by Electronic Publishing Services, Inc. The editors were Barbara A. Bodling and Frances Tindall.
Braun-Brumfield, Inc. was printer and binder.

Library of Congress Cataloging in Publication Data

Kraus, Allan D.
  Matrices for engineers.

  Bibliography: p.
  Includes index.
  1. Matrices. I. Title.
QA188.K73   1987   512.9′434   87-139
ISBN 0-89116-646-7   Hemisphere Publishing Corporation

DISTRIBUTION OUTSIDE NORTH AMERICA:
ISBN 3-540-17785-X Springer-Verlag Berlin

To my children of whom I am very proud.
Deborah, Raymond, Lawrence, and Edward.

CONTENTS

6 ORTHOGONALITY AND COORDINATE TRANSFORMATIONS

7 THE EIGENVALUE PROBLEM

PREFACE

In teaching elementary network theory, engineering mechanics, and vibration analysis, I have found that many students lack versatility in the manipulation of matrices and little or no foundation in the theory of matrices. This book attempts to alleviate this situation, either in a self-study or classroom framework. The approach here is to provide the necessary material in a direct manner, in most cases without rigorous proofs and derivations, because it is believed that the proof is often formidable and tends to obstruct, rather than aid, the learning process. Numerous examples are provided to illustrate each concept as it appears.

The emphasis is on methodology rather than rigor, and those who are looking for rigor may not find enough of it here to satisfy them. The book considers such topics as linear transformations from an intuitive and engineering point of view that conditions at one point in a system induce conditions at another. Little is included on vector spaces and bases; however, when a proof is interesting and instructive, it is included. And, to be sure, an eye has been kept on the use of computational techniques.

Moreover, the book is more than a book on matrix algebra. Matrix calculus is included as well, as are real world engineering examples. Also included is a summary of methods for finding the roots of polynomial equations.

The author is pleased to acknowledge the assistance of his wife Ruth, who, as usual, assisted ably with the production process.

Allan D. Kraus

MATRICES FOR ENGINEERS

ONE

PRELIMINARY CONCEPTS

1.1 INTRODUCTION

Before the advent of the digital computer, the hand-held calculator, and the desk-top computer, one could talk about the solution of large-scale systems of linear algebraic equations. But few had ever attempted a solution of such a system, and those who had made the attempt were confronted with a most laborious and detailed computational procedure. Now, as a result of the technological revolution that began in the late 1940s and early 1950s, large systems of n algebraic equations in n unknowns (with n in the hundreds and even the thousands) can be formed expeditiously and solved efficiently using the modern, high-speed, digital computer.

The formulation and solution of such equations relies on an ordered methodology that is matrix-oriented. But the use of matrices and matrix theory is not restricted to the solution of systems of algebraic equations. There are many applications in pure and applied mathematics where one is confronted with rectangular arrays of variables and numbers. Matrix methods are employed in such diverse analysis domains as structural analysis, the theory of elasticity, classical mechanics, electrical network analysis, control system synthesis, and the analysis of mechanical vibration problems.

1.2 BASIC CONCEPTS

A system of n linear algebraic equations in n unknowns $x_1, x_2, x_3, \ldots,$ x_n such as

$$a_{11} x_1 + a_{12} x_2 + a_{13} x_3 + \cdots a_{1n}x_n = y_1$$

$$a_{21} x_1 + a_{22} x_2 + a_{23} x_3 + \cdots a_{2n}x_n = y_2$$

$$a_{31} x_1 + a_{32} x_2 + a_{33} x_3 + \cdots a_{3n}x_n = y_3$$

$$\cdots$$

$$a_{n1} x_1 + a_{n2} x_2 + a_{n3} x_3 + \cdots a_{nn}x_n = y_n$$

can conveniently be written as a matrix equation in the form

$$
\begin{bmatrix}
a_{11} & a_{12} & a_{13} & \cdots & a_{1n} \\
a_{21} & a_{22} & a_{23} & \cdots & a_{2n} \\
a_{31} & a_{32} & a_{33} & \cdots & a_{3n} \\
& & \cdots & & \\
a_{n1} & a_{n2} & a_{n3} & \cdots & a_{nn}
\end{bmatrix}
\begin{bmatrix}
x_1 \\
x_2 \\
x_3 \\
\cdots \\
x_n
\end{bmatrix}
=
\begin{bmatrix}
y_1 \\
y_2 \\
y_3 \\
\cdots \\
y_n
\end{bmatrix}
$$

or more simply by

$$AX = Y$$

where A is a rectangular matrix (in this case square) having elements a_{ij} and where X and Y are column vectors with elements x_j and y_i, respectively. The foregoing representations imply that

$$\sum_{j=1}^{n} a_{ij} x_j = y_i \qquad i = 1, 2, 3, \cdots, n$$

The matrix A is called the *coefficient matrix*. If it is desired to associate the elements of Y with the coefficient matrix A, one may augment A and define an augmented matrix

$$
A^a =
\begin{bmatrix}
a_{11} & a_{12} & a_{13} & \cdots & a_{1n} & y_1 \\
a_{21} & a_{22} & a_{23} & \cdots & a_{2n} & y_2 \\
a_{31} & a_{32} & a_{33} & \cdots & a_{3n} & y_3 \\
& & \cdots & & & \\
a_{n1} & a_{n2} & a_{n3} & \cdots & a_{nn} & y_n
\end{bmatrix}
$$

which has n rows and $n + 1$ columns. This matrix may be written more simply as the augmented matrix

$$A^a = [A{:}Y]$$

where the superscript means *augmented* and where the idea of a *partitioned matrix* is apparent.

Example: In the system of linear algebraic equations

$$8x_1 - 3x_2 + 2x_3 = 14$$
$$-3x_1 + 7x_2 - x_3 = -3$$
$$x_1 - 6x_2 + 11x_3 = 17$$

the matrix

$$\begin{bmatrix} 8 & -3 & 2 \\ -3 & 7 & -1 \\ 1 & -6 & 11 \end{bmatrix}$$

is the *coefficient matrix A* of the system

$$AX = B$$

and the matrix

$$\begin{bmatrix} 8 & -3 & 2 & 14 \\ -3 & 7 & -1 & -3 \\ 1 & -6 & 11 & 17 \end{bmatrix}$$

that contains the constant terms, in addition to the elements of A, is the *augmented matrix* of the system. Moreover, the unknowns and the constant terms form two column vectors X and B.

In the representation

$$AX = Y$$

A is said to premultiply X (A is a premultiplier), and X is said to post-multiply A (X is a postmultiplier).

1.3 MATRIX AND VECTOR TERMINOLOGY

A matrix of order $m \times n$

$$\begin{bmatrix} a_{11} & a_{12} & a_{13} & \cdots & a_{1n} \\ a_{21} & a_{22} & a_{23} & \cdots & a_{2n} \\ a_{31} & a_{32} & a_{33} & \cdots & a_{3n} \\ & & \cdots & & \\ a_{m1} & a_{m2} & a_{m3} & \cdots & a_{mn} \end{bmatrix}$$

is a rectangular-ordered array of a total of mn entries arranged in m rows and n columns. The order of this matrix is $m \times n$, which is often written as (m,n).

If $m = n$, the matrix is square of order $n \times n$ (or of n or of nth order)

$$\begin{bmatrix} a_{11} & a_{12} & a_{13} & \cdots & a_{1n} \\ a_{21} & a_{22} & a_{23} & \cdots & a_{2n} \\ a_{31} & a_{32} & a_{33} & \cdots & a_{3n} \\ & & \cdots & & \\ a_{n1} & a_{n2} & a_{n3} & \cdots & a_{nn} \end{bmatrix}$$

In both the rectangular and square matrices, a_{ij} is called the ij element of A, and if the matrix is square and $i = j$, the element is said to define and be located on the principal diagonal. The elements a_{n1}, $a_{n-1,2}$, $a_{n-2,3}$, $\cdots$, a_{1n} are located on and constitute the secondary diagonal.

All elements where $i \neq j$ are considered to be off-diagonal, subdiagonal if $i > j$, and superdiagonal if $i < j$. The sum of the elements on the principal diagonal of A is called the *trace* of A

$$\text{tr } A = \sum_{k=1}^{n} a_{kk}$$

Example: The matrix

$$A = \begin{bmatrix} 4 & 2 & 0 & 1 \\ -3 & 4 & 8 & 0 \\ -1 & -1 & 6 & 2 \\ -2 & 7 & 1 & 9 \end{bmatrix}$$

is square and is of fourth order (4×4). The elements 4, 4, 6, and 9 constitute the principal diagonal, and the elements -2, -1, 8, and 1 constitute the secondary diagonal. The element 8 is the a_{23} element, which lies at the intersection of the second row and third column.

The trace of A is

$$\text{tr } A = 4 + 4 + 6 + 9 = 23$$

A vector is a matrix containing a single row or a single column. If it is a $1 \times n$ matrix (a matrix of order $1 \times n$), it is a row vector

$$V = [v_1 \ v_2 \ v_3 \cdots v_n]$$

If the vector is an $m \times 1$ vector (order $m \times 1$), it is a column vector

$$V = \begin{bmatrix} v_1 \\ v_2 \\ v_3 \\ \cdots \\ v_m \end{bmatrix}$$

which some authors write with braces as

$$V = \{v_1 \, v_2 \, v_3 \cdots v_m\}$$

to conserve space. This procedure, because it may tend to confuse the reader with notation in set theory, will not be followed in this book.

This concept and the usual one regarding a vector (in vector analysis, a vector is defined as a quantity having magnitude and direction) have certain similarities which will be developed as this study proceeds. Because of these similarities, the elements of a vector are frequently called components. However, caution is necessary, because the usual three-dimensional space in vector analysis does not indicate that m or n (for column or row vectors, respectively) is limited to an upper bound of 3.

Moreover, the order of a matrix should not be confused with the dimension of a matrix, which is a term that comes forth in the study of vector spaces.

1.4 SOME SPECIAL MATRICES

An $m \times n$ matrix such as the one displayed in the previous section is called a null matrix if every element in the matrix is identically equal to zero. For example, the 2×3 matrix

$$\begin{bmatrix} 0 & 0 & 0 \\ 0 & 0 & 0 \end{bmatrix}$$

is null.

The transpose of an $m \times n$ matrix is an $n \times m$ matrix with the rows and columns of the original matrix interchanged. For the 3×4 matrix

$$A = \begin{bmatrix} 3 & 4 & -2 & 1 \\ 2 & 2 & 1 & 0 \\ 1 & -4 & -3 & 1 \end{bmatrix}$$

the transpose is 4×3

$$A^{\mathrm{T}} = \begin{bmatrix} 3 & 2 & 1 \\ 4 & 2 & -4 \\ -2 & 1 & -3 \\ 1 & 0 & 1 \end{bmatrix}$$

Note the use of the superscript T to indicate the transpose and recognize that the transpose of the transpose is the original matrix

$$[A^{\mathrm{T}}]^{\mathrm{T}} = A$$

The foregoing remarks concerning the null matrix and the transpose pertain as well to square matrices. However, in addition to these, there are several special square matrices. Just a few of these are presented in this section.

Consider the nth-order square matrix

$$\begin{bmatrix} a_{11} & a_{12} & a_{13} & \cdots & a_{1n} \\ a_{21} & a_{22} & a_{23} & \cdots & a_{2n} \\ a_{31} & a_{32} & a_{33} & \cdots & a_{3n} \\ & & \cdots & & \\ a_{n1} & a_{n2} & a_{n3} & \cdots & a_{nn} \end{bmatrix}$$

If the elements $a_{ij} = 0$ for all $i \neq j$, the matrix is said to be diagonal

$$A = \begin{bmatrix} a_{11} & 0 & 0 & \cdots & 0 \\ 0 & a_{22} & 0 & \cdots & 0 \\ 0 & 0 & a_{33} & \cdots & 0 \\ & & \cdots & & \\ 0 & 0 & 0 & \cdots & a_{nn} \end{bmatrix}$$

and if all a_{ij} are equal for $i = j$ (that is, $a_{ij} = \alpha$; $i = j$ and $a_{ij} = 0$; $i \neq j$), then the resulting matrix is said to be a *scalar matrix*, which is a diagonal matrix with all principal diagonal elements equal.

$$A = \begin{bmatrix} \alpha & 0 & 0 & \cdots & 0 \\ 0 & \alpha & 0 & \cdots & 0 \\ 0 & 0 & \alpha & \cdots & 0 \\ & & \cdots & & \\ 0 & 0 & 0 & \cdots & \alpha \end{bmatrix}$$

Finally, if all α in the scalar matrix are equal to unity ($\alpha = 1$), the scalar matrix becomes the identity matrix

$$I = \begin{bmatrix} 1 & 0 & 0 & \cdots & 0 \\ 0 & 1 & 0 & \cdots & 0 \\ 0 & 0 & 1 & \cdots & 0 \\ & & \cdots & & \\ 0 & 0 & 0 & \cdots & 1 \end{bmatrix}$$

Many more special square matrices are considered in a later section in this chapter.

1.5 MATRIX EQUALITY

A matrix $A = [a_{ij}]_{m \times n}$ is equal to a matrix $B = [b_{ij}]_{m \times n}$ if and only if $a_{ij} = b_{ij}$ for all i and j. This essentially states that two matrices are equal if and only if they are of the same order and corresponding elements are equal.

1.6 MATRIX ADDITION AND SUBTRACTION

A matrix $A = [a_{ij}]_{m \times n}$ may be added to a matrix $B = [b_{ij}]_{m \times n}$ to form a matrix $C = [c_{ij}]_{m \times n} = [a_{ij} + b_{ij}]_{m \times n}$. This points out that to form the sum of two matrices, the matrices must be of the same order (in this event, the matrices are said to be conformable for addition) and that the elements of the sum are determined by adding the corresponding elements of the matrices forming the sum.

Example: if

$$A = \begin{bmatrix} 1 & 2 & 3 \\ 4 & 5 & 6 \end{bmatrix}$$

$$B = \begin{bmatrix} -1 & 0 & 4 \\ 3 & -2 & -4 \end{bmatrix}$$

and

$$C = \begin{bmatrix} 2 & 1 \\ 3 & -1 \end{bmatrix}$$

then

$$A + B = \begin{bmatrix} (1 - 1) & (2 + 0) & (3 + 4) \\ (4 + 3) & (5 - 2) & (6 - 4) \end{bmatrix} = \begin{bmatrix} 0 & 2 & 7 \\ 7 & 3 & 2 \end{bmatrix}$$

and $A + C$ and $B + C$ do not exist, because the order of C differs from the orders of both A and B, that is, A and B are not comfortable with C for addition.

Matrix addition is both commutative and associative:

$$A + B = B + A$$

$$A + (B + C) = (A + B) + C$$

In addition, the sum $A + C$ is equal to the sum $B + C$ if and only if $A = B$. This is the cancellation law for addition.

The matrix $B = [b_{ij}]_{m \times n}$ may be subtracted from the matrix $A = [a_{ij}]_{m \times n}$ to form the matrix $D = [d_{ij}]_{m \times n} = [a_{ij} - b_{ij}]_{m \times n}$. This indicates that two matrices of the same order and conformable for subtraction may be subtracted by forming the difference between the corresponding elements of the minuend and the subtrahend. Moreover, it is easy to see that if

$$A + B = C$$

then

$$A = C - B$$

Finally, it may be observed that a square matrix possesses a unique decomposition into a sum of a subdiagonal, a diagonal, and a superdiagonal matrix. For example

$$A = \begin{bmatrix} 1 & 2 & 3 & 4 \\ 5 & 6 & 7 & 8 \\ 5 & 6 & 7 & 8 \\ 4 & 3 & 2 & 1 \end{bmatrix} = \begin{bmatrix} 0 & 0 & 0 & 0 \\ 5 & 0 & 0 & 0 \\ 5 & 6 & 0 & 0 \\ 4 & 3 & 2 & 0 \end{bmatrix} + \begin{bmatrix} 1 & 0 & 0 & 0 \\ 0 & 6 & 0 & 0 \\ 0 & 0 & 7 & 0 \\ 0 & 0 & 0 & 1 \end{bmatrix} + \begin{bmatrix} 0 & 2 & 3 & 4 \\ 0 & 0 & 7 & 8 \\ 0 & 0 & 0 & 8 \\ 0 & 0 & 0 & 0 \end{bmatrix}$$

1.7 MATRIX MULTIPLICATION

A matrix may be multiplied by a scalar or by another matrix. If $A = [a_{ij}]$ and α is a scalar, then

$$\alpha A = A \alpha = [\alpha a_{ij}]$$

This shows that multiplication by a scalar is commutative and that multiplication by a scalar involves the multiplication of each and every element of the matrix by the scalar. In addition, it is easy to see that

$$(\alpha + \beta)A = \alpha A + \beta A$$
$$\alpha(A + B) = \alpha A + \alpha B$$

and

$$\alpha(\beta A) = (\alpha\beta)A$$

Observe that a scalar matrix is equal to the product of the scalar and the identity matrix. For example,

$$\begin{bmatrix} 3 & 0 & 0 \\ 0 & 3 & 0 \\ 0 & 0 & 3 \end{bmatrix} = 3 \begin{bmatrix} 1 & 0 & 0 \\ 0 & 1 & 0 \\ 0 & 0 & 1 \end{bmatrix}$$

A modest effort must be expended to use the terminology *multiplication by a scalar* to avoid confusion with the process known as *scalar multiplication*.

The product of a row matrix of order $1 \times n$ and a column matrix of order $n \times 1$ forms a 1×1 matrix, which has no important property that is not possessed by a scalar. This product is therefore called the scalar or dot product (some sources also use the terminology *inner product*). It is denoted by a dot placed between the two matrices in the product, that is,

$$A \cdot B = [a_{1j}]_{1 \times n} \cdot [b_{j1}]_{n \times 1} = \gamma$$

where A and B are column vectors and where γ is a scalar obtained from

$$\gamma = \sum_{k=1}^{n} a_k b_k$$

If the scalar product of two vectors is uniquely equal to zero, then the vectors are said to be *orthogonal*. A more detailed treatment of the important concept of orthogonality is provided in Chapter 6.

Example: If

$$A = \begin{bmatrix} 1 & 2 & 3 & 4 & 2 \end{bmatrix}$$

and

$$B = \begin{bmatrix} 4 \\ -5 \\ -3 \\ -2 \\ 7 \end{bmatrix}$$

then

$$A \cdot B = 1(4) + 2(-5) + 3(-3) + 4(-2) + 2(7)$$
$$= 4 - 10 - 9 - 8 + 14 = -9$$

Example: If

$$A = [1 \ \ 2 \ \ 3 \ \ 4 \ \ 4]$$

the value of x in

$$B = \begin{bmatrix} 1 \\ -3 \\ 4 \\ x \\ -1 \end{bmatrix}$$

that makes A orthogonal to B is found through the scalar or dot product set equal to zero:

$$A \cdot B = 1(1) + 2(-3) + 3(4) + 4x + 4(-1) = 0$$
$$= 1 - 6 + 12 + 4x - 4 = 0$$

Then

$$4x = -3$$

and

$$x = -3/4$$

In Section 1.2 a set of linear simultaneous algebraic equations has been shown to be represented by the notation

$$AX = Y$$

where A is the $n \times n$ coefficient matrix and X and Y are $n \times 1$ column vectors. To obtain the original set of equations from a set where $n = 3$

$$\begin{bmatrix} a_{11} & a_{12} & a_{13} \\ a_{21} & a_{22} & a_{23} \\ a_{31} & a_{32} & a_{33} \end{bmatrix} \begin{bmatrix} x_1 \\ x_2 \\ x_3 \end{bmatrix} = \begin{bmatrix} y_1 \\ y_2 \\ y_3 \end{bmatrix}$$

a row by column element product and sum operation is clearly evident:

$$a_{11}x_1 + a_{12}x_2 + a_{13}x_3 = y_1$$

$$a_{21}x_1 + a_{22}x_2 + a_{23}x_3 = y_2$$

$$a_{31}x_1 + a_{32}x_2 + a_{33}x_3 = y_3$$

and it is observed that each element of y is obtained by multiplying the corresponding elements of A by the elements of X and adding the results.

Notice that the foregoing procedure is not possible if the number of columns of A does not equal the number of rows of X. In this event there are not always corresponding elements to multiply. Moreover, it should be noted that Y contains the same number of rows as both A and X.

The foregoing suggests a general definition for the multiplication of two matrices. If A is $m \times n$ and B is $p \times q$, $AB = C$ exists if $n = p$, in which case the matrix C is $m \times q$ with elements given by

$$[c_{ij}]_{m \times q} = \sum_{k=1}^{n=p} a_{ik}b_{kj} \qquad \begin{aligned} i &= 1, 2, 3, \cdots, m \\ j &= 1, 2, 3, \cdots, q \end{aligned}$$

When $n = p$, the matrices A and B are said to be conformable for multiplication.

Example: If

$$A = \begin{bmatrix} 1 & 2 & 3 & 4 \\ -1 & 3 & -2 & 0 \end{bmatrix}$$

$$B = A^{\mathrm{T}} = \begin{bmatrix} 1 & -1 \\ 2 & 3 \\ 3 & -2 \\ 4 & 0 \end{bmatrix}$$

and

$$C = \begin{bmatrix} 1 & 2 \\ 3 & 4 \end{bmatrix}$$

then the product AB exists because A is 2×4 and B is 4×2. The result AB is 2×2

$$AB = \begin{bmatrix} 1 & 2 & 3 & 4 \\ -1 & 3 & -2 & 0 \end{bmatrix} \begin{bmatrix} 1 & -1 \\ 2 & 3 \\ 3 & -2 \\ 4 & 0 \end{bmatrix}$$

$$= \begin{bmatrix} (1 + 4 + 9 + 16) & (-1 + 6 - 6 + 0) \\ (-1 + 6 - 6 + 0) & (1 + 9 + 4 + 0) \end{bmatrix}$$

or

$$AB = \begin{bmatrix} 30 & -1 \\ -1 & 14 \end{bmatrix}$$

The product BA also exists:

$$BA = \begin{bmatrix} 1 & -1 \\ 2 & 3 \\ 3 & -2 \\ 4 & 0 \end{bmatrix} \begin{bmatrix} 1 & 2 & 3 & 4 \\ -1 & 3 & -2 & 0 \end{bmatrix}$$

$$= \begin{bmatrix} (1+1) & (2-3) & (3+2) & (4+0) \\ (2-3) & (4+9) & (6-6) & (8+0) \\ (3+2) & (6-6) & (9+4) & (12+0) \\ (4+0) & (8+0) & (12+0) & (16+0) \end{bmatrix}$$

or

$$BA = \begin{bmatrix} 2 & -1 & 5 & 4 \\ -1 & 13 & 0 & 8 \\ 5 & 0 & 13 & 12 \\ 4 & 8 & 12 & 16 \end{bmatrix}$$

Notice that $AB \neq BA$, which shows that matrix multiplication, in general, is not commutative. Notice also that the product CA exists because C is 2×2 and A is 2×4. The product is 2×4:

$$CA = \begin{bmatrix} 1 & 2 \\ 3 & 4 \end{bmatrix} \begin{bmatrix} 1 & 2 & 3 & 4 \\ -1 & 3 & -2 & 0 \end{bmatrix} = \begin{bmatrix} -1 & 8 & -1 & 4 \\ -1 & 18 & 1 & 12 \end{bmatrix}$$

and it is easy to see from the conformability rules that the products AC and CB do not exist.

However,

$$BC = \begin{bmatrix} 1 & -1 \\ 2 & 3 \\ 3 & -2 \\ 4 & 0 \end{bmatrix} \begin{bmatrix} 1 & 2 \\ 3 & 4 \end{bmatrix} = \begin{bmatrix} -2 & -2 \\ 11 & 16 \\ -3 & -2 \\ 4 & 8 \end{bmatrix}$$

It has been demonstrated that, in general, the commutative law does not hold. That is, except in special cases

$$AB \neq BA$$

However, it can be proved with some difficulty that the multiplication of matrices is associative:

$$(AB)C = A(BC)$$

and matrix multiplication is distributive with respect to addition:

$$A(B + C) = AB + AC$$
$$(A + B)C = AC + BC$$

assuming that conformability exists for both addition and multiplication.

Example: If

$$A = \begin{bmatrix} 1 & 2 \\ -1 & 0 \end{bmatrix}$$

$$B = \begin{bmatrix} 4 & -1 \\ -1 & 2 \end{bmatrix}$$

and

$$C = \begin{bmatrix} 1 & 0 \\ -2 & 3 \end{bmatrix}$$

then

$$AB = \begin{bmatrix} 1 & 2 \\ -1 & 0 \end{bmatrix} \begin{bmatrix} 4 & -1 \\ -1 & 2 \end{bmatrix} = \begin{bmatrix} 2 & 3 \\ -4 & 1 \end{bmatrix}$$

and

$$(AB)C = \begin{bmatrix} 2 & 3 \\ -4 & 1 \end{bmatrix} \begin{bmatrix} 1 & 0 \\ -2 & 3 \end{bmatrix} = \begin{bmatrix} -4 & 9 \\ -6 & 3 \end{bmatrix}$$

Moreover,

$$BC = \begin{bmatrix} 4 & -1 \\ -1 & 2 \end{bmatrix} \begin{bmatrix} 1 & 0 \\ -2 & 3 \end{bmatrix} = \begin{bmatrix} 6 & -3 \\ -5 & 6 \end{bmatrix}$$

and

$$A(BC) = \begin{bmatrix} 1 & 2 \\ -1 & 0 \end{bmatrix} \begin{bmatrix} 6 & -3 \\ -5 & 6 \end{bmatrix} = \begin{bmatrix} -4 & 9 \\ -6 & 3 \end{bmatrix}$$

which shows that

$$(AB)C = A(BC)$$

Now

$$AB = \begin{bmatrix} 2 & 3 \\ -4 & 1 \end{bmatrix}$$

and

$$AC = \begin{bmatrix} 1 & 2 \\ -1 & 0 \end{bmatrix} \begin{bmatrix} 1 & 0 \\ -2 & 3 \end{bmatrix} = \begin{bmatrix} -3 & 6 \\ -1 & 0 \end{bmatrix}$$

so that

$$AB + AC = \begin{bmatrix} 2 & 3 \\ -4 & 1 \end{bmatrix} + \begin{bmatrix} -3 & 6 \\ -1 & 0 \end{bmatrix} = \begin{bmatrix} -1 & 9 \\ -5 & 1 \end{bmatrix}$$

With

$$B + C = \begin{bmatrix} 4 & -1 \\ -1 & 2 \end{bmatrix} + \begin{bmatrix} 1 & 0 \\ -2 & 3 \end{bmatrix} = \begin{bmatrix} 5 & -1 \\ -3 & 5 \end{bmatrix}$$

it is seen that

$$A(B + C) = \begin{bmatrix} 1 & 2 \\ -1 & 0 \end{bmatrix} \begin{bmatrix} 5 & -1 \\ -3 & 5 \end{bmatrix} = \begin{bmatrix} -1 & 9 \\ -5 & 1 \end{bmatrix}$$

which shows that

$$A(B + C) = AB + AC$$

Finally, with

$$AC = \begin{bmatrix} -3 & 6 \\ -1 & 0 \end{bmatrix}$$

and

$$BC = \begin{bmatrix} 6 & -3 \\ -5 & 6 \end{bmatrix}$$

then

$$AC + BC = \begin{bmatrix} -3 & 6 \\ -1 & 0 \end{bmatrix} + \begin{bmatrix} 6 & -3 \\ -5 & 6 \end{bmatrix} = \begin{bmatrix} 3 & 3 \\ -6 & 6 \end{bmatrix}$$

and with

$$A + B = \begin{bmatrix} 1 & 2 \\ -1 & 0 \end{bmatrix} + \begin{bmatrix} 4 & -1 \\ -1 & 2 \end{bmatrix} = \begin{bmatrix} 5 & 1 \\ -2 & 2 \end{bmatrix}$$

then

$$(A + B)C = \begin{bmatrix} 5 & 1 \\ -2 & 2 \end{bmatrix} \begin{bmatrix} 1 & 0 \\ -2 & 3 \end{bmatrix} = \begin{bmatrix} 3 & 3 \\ -6 & 6 \end{bmatrix}$$

which confirms that

$$(A + B)C = AC + BC$$

If the product AB is null, that is, $AB = 0$, it cannot be concluded that either A or B is null. Furthermore, if $AB = AC$ or $CA = BA$, it cannot be concluded that $B = C$. This means that, in general, cancellation of matrices is not permissible.

Example: If

$$A = \begin{bmatrix} 1 & 2 \\ -1 & 3 \end{bmatrix}$$

$$B = \begin{bmatrix} 4 & -1 \\ 0 & 0 \end{bmatrix}$$

and

$$C = \begin{bmatrix} 1 & 3 \\ -1 & 7 \end{bmatrix}$$

then

$$AB = \begin{bmatrix} 1 & 2 \\ -1 & 3 \end{bmatrix} \begin{bmatrix} 4 & -1 \\ 0 & 0 \end{bmatrix} = \begin{bmatrix} 4 & -1 \\ -4 & 1 \end{bmatrix}$$

and

$$CB = \begin{bmatrix} 1 & 3 \\ -1 & 7 \end{bmatrix} \begin{bmatrix} 4 & -1 \\ 0 & 0 \end{bmatrix} = \begin{bmatrix} 4 & -1 \\ -4 & 1 \end{bmatrix}$$

which shows that

$$AB = CB$$

even though $A \neq C$

Example: If

$$A = \begin{bmatrix} 0 & 1 & 2 \\ 0 & -1 & 2 \\ 0 & 3 & 5 \end{bmatrix}$$

and

$$B = \begin{bmatrix} 3 & 2 & -1 \\ 0 & 0 & 0 \\ 0 & 0 & 0 \end{bmatrix}$$

it is easy to see that

$$AB = \begin{bmatrix} 0 & 1 & 2 \\ 0 & -1 & 2 \\ 0 & 3 & 5 \end{bmatrix} \begin{bmatrix} 3 & 2 & -1 \\ 0 & 0 & 0 \\ 0 & 0 & 0 \end{bmatrix} = \begin{bmatrix} 0 & 0 & 0 \\ 0 & 0 & 0 \\ 0 & 0 & 0 \end{bmatrix}$$

Thus the product AB yields the 3×3 null matrix, even though neither A nor B is null.

1.8 MATRIX DIVISION AND MATRIX INVERSION

Matrix division is not defined. Instead, use is made of a process called *matrix inversion* which relies on the existence of the identity matrix that is related to a square matrix A by

$$AI = IA = A$$

Consider the identity for addition, 0, that has the property that for all scalars, α

$$\alpha + 0 = 0 + \alpha = \alpha$$

and an identity element for multiplication 1 so that

$$\alpha 1 = 1\alpha = \alpha$$

The scalar most certainly possesses a reciprocal or multiplicative inverse $1/\alpha$ that when multiplied by α yields the identity element for scalar multiplication:

$$(1/\alpha) = (\alpha^{-1})\alpha = 1$$

This reasoning may be extended to the $n \times n$ matrix A and the pair of identity matrices; the $n \times n$ identity matrix for multiplication I and the $n \times n$ identity matrix for addition 0 (a null matrix). Thus, as already noted,

$$AI = IA = A$$

and

$$A + 0 = 0 + A = A$$

If there is an $n \times n$ matrix A^{-1} that pre- and postmultiplies A such that

$$A^{-1}A = AA^{-1} = I$$

then A^{-1} is the inverse of A with respect to matrix multiplication. The matrix A is said to be invertible or nonsingular if A^{-1} exists and noninvertible or singular if A^{-1} does not exist.

Example: The 3 × 3 matrix

$$A = \begin{bmatrix} 3 & -2 & 0 \\ -2 & 4 & -1 \\ 0 & -1 & 6 \end{bmatrix}$$

can be shown to possess the inverse (Chapter 3)

$$A^{-1} = \begin{bmatrix} 23/45 & 4/15 & 2/45 \\ 4/15 & 2/5 & 1/15 \\ 2/45 & 1/15 & 8/45 \end{bmatrix}$$

A simple multiplication produces the identity matrix:

$$\begin{bmatrix} 3 & -2 & 0 \\ -2 & 4 & -1 \\ 0 & -1 & 6 \end{bmatrix} \begin{bmatrix} 23/45 & 4/15 & 2/45 \\ 4/15 & 2/5 & 1/15 \\ 2/45 & 1/15 & 8/45 \end{bmatrix} = \begin{bmatrix} 1 & 0 & 0 \\ 0 & 1 & 0 \\ 0 & 0 & 1 \end{bmatrix}$$

One can also go the other way:

$$\begin{bmatrix} 23/45 & 4/15 & 2/45 \\ 4/15 & 2/5 & 1/15 \\ 2/45 & 1/15 & 8/45 \end{bmatrix} \begin{bmatrix} 3 & -2 & 0 \\ -2 & 4 & -1 \\ 0 & -1 & 6 \end{bmatrix} = \begin{bmatrix} 1 & 0 & 0 \\ 0 & 1 & 0 \\ 0 & 0 & 1 \end{bmatrix}$$

The concept of matrix inversion is of paramount importance to the study of matrix methods. For this reason, Chapter 3 is devoted in its entirety to a detailed consideration of this subject.

1.9 POWERS AND ROOTS OF SQUARE MATRICES

Because an $n \times n$ matrix, A, is conformable with itself for multiplication, one may form its powers

$$A^n = AAA \cdots A$$

In addition, it is easy to see that the law of exponents holds

$$A^m A^n = A^{(m+n)}$$

and the zeroth power of the matrix A is the identity matrix.

Negative powers may be defined:

$$A^{-n} = (A^{-1})^n$$

and so may roots:

$$A^{1/n} = \sqrt[n]{A}$$

Further discussion of these ideas is presented at appropriate places as this study proceeds.

1.10 FURTHER PROPERTIES OF THE TRANSPOSE

Recall that a matrix A has a transpose A^T where the rows of A are the columns of A^T and the columns of A are the rows of A^T. It is easy to see that the transpose of the transpose is the matrix itself:

$$[A^T]^T = A$$

and in accordance with the rules for matrix addition and multiplication by a scalar developed in Sections 1.6 and 1.7:

$$(A + B + C + \cdots)^T = A^T + B^T + C^T \cdots$$

$$(\alpha A)^T = \alpha A^T$$

and

$$(A^n)^T = (A^T)^n$$

The determination of the transpose of the product of two matrices presents an interesting exercise in matrix algebra. Consider two matrices A and B and their product $C = AB$. Then the ijth element of $C^T = (AB)^T$ is the jith element of C. Because individual elements are obtained from row-by-column products and sums, the jith element of C can be seen to derive from the dot product

$$[c_{ji}] = [j\text{th row of } A] \cdot [i\text{th column of } B]$$

or

$$[c_{ji}] = (a_{j1} \ a_{j2} \ a_{j3} \cdots a_{jk}) \cdot (b_{1i} \ b_{2i} \ b_{3i} \cdots b_{ki})$$

But the dot product is commutative, so that

$$[c_{ji}] = (b_{1i} \ b_{2i} \ b_{3i} \cdots b_{ki}) (a_{j1} \ a_{j2} \ a_{j3} \cdots a_{jk})$$

or

$$[c_{ji}] = [i\text{th row of } B^T] \cdot [j\text{th column of } A^T]$$

This is the ijth element of $B^T A^T$, and because $(AB)^T$ has the same order as $B^T A^T$,

$$(AB)^T = B^T A^T$$

which shows that the transpose of the product of two matrices is equal to the product of the individual transposes taken in reverse order.

Example: If

$$A = \begin{bmatrix} 1 & 2 & 3 \\ 4 & -1 & -2 \\ 0 & -2 & 1 \\ 2 & -3 & 2 \end{bmatrix}$$

so that

$$A^\mathrm{T} = \begin{bmatrix} 1 & 4 & 0 & 2 \\ 2 & -1 & -2 & -3 \\ 3 & -2 & 1 & 2 \end{bmatrix}$$

and

$$B = \begin{bmatrix} 1 & 1 \\ 2 & -1 \\ 0 & 3 \end{bmatrix}$$

so that

$$B^\mathrm{T} = \begin{bmatrix} 1 & 2 & 0 \\ 1 & -1 & 3 \end{bmatrix}$$

then

$$AB = \begin{bmatrix} 1 & 2 & 3 \\ 4 & -1 & -2 \\ 0 & -2 & 1 \\ 2 & -3 & 2 \end{bmatrix} \begin{bmatrix} 1 & 1 \\ 2 & -1 \\ 0 & 3 \end{bmatrix} = \begin{bmatrix} 5 & 8 \\ 2 & -1 \\ -4 & 5 \\ -4 & 11 \end{bmatrix}$$

and

$$(AB)^\mathrm{T} = \begin{bmatrix} 5 & 2 & -4 & -4 \\ 8 & -1 & 5 & 11 \end{bmatrix}$$

Now find the product

$$(B^\mathrm{T}A^\mathrm{T}) = \begin{bmatrix} 1 & 2 & 0 \\ 1 & -1 & 3 \end{bmatrix} \begin{bmatrix} 1 & 4 & 0 & 2 \\ 2 & -1 & -2 & -3 \\ 3 & -2 & 1 & 2 \end{bmatrix} = \begin{bmatrix} 5 & 2 & -4 & -4 \\ 8 & -1 & 5 & 11 \end{bmatrix}$$

which shows that $(AB)^\mathrm{T} = B^\mathrm{T}A^\mathrm{T}$.

1.11 SOME SPECIAL MATRICES

There are several types of matrices that possess unique properties. Because they frequently appear in engineering applications, they are listed here.

The diagonal, scalar, and identity matrices are square matrices which were introduced in Section 1.4. A square matrix A is said to be symmetric if for all $i \neq j$

$$[a_{ij}] = [a_{ji}]$$

Notice that this requires that

$$A = A^T \qquad \text{(symmetric matrix)}$$

A square matrix A is said to be a skew matrix if all the principal diagonal elements are not zero and for all $i \neq j$

$$[a_{ij}] = -[a_{ji}] \qquad \text{(skew matrix)}$$

This matrix becomes skew-symmetric if all the principal diagonal elements are equal to zero. In this case

$$A = -A^T \qquad \text{(skew-symmetric matrix)}$$

Any matrix may be composed of complex elements

$$C = [c_{ij}] = [a_{ij} + jb_{ij}] \qquad \text{(complex matrix)}$$

where $j = (-1)^{1/2}$. Such a matrix has a conjugate

$$\overline{C} = [\overline{c}_{ij}] = [a_{ij} - jb_{ij}] \qquad \text{(conjugate matrix)}$$

and if C is composed of real elements, it can be called a real matrix, in which case

$$C = \overline{C} \qquad \text{(real matrix)}$$

If C is composed of purely imaginary elements, it is designated as pure imaginary, and

$$C = -\overline{C} \qquad \text{(imaginary matrix)}$$

If the matrix C is square and consists of only real numbers for all $i = j$, it is called Hermitian and exhibits the property that

$$C = \overline{C}^T \qquad \text{(Hermitian matrix)}$$

This matrix is skew-Hermitian if

$$C = -\overline{C}^T \qquad \text{(skew-Hermitian matrix)}$$

and if the matrix is equal to the inverse of the transpose of the conjugate matrix, the matrix is said to be unitary

$$C = [\overline{C}^T]^{-1} \qquad \text{(unitary matrix)}$$

Finally, if a square matrix is invertible and equal to its inverse, it is said to be involutory

$$C = C^{-1}$$ (involutory matrix)

1.12 EXERCISES

Exercises 1.1 to 1.16 are based on a consideration of the matrices A through G.

$$A = \begin{bmatrix} 1 & 2 & -1 \\ 3 & 4 & -2 \end{bmatrix} \qquad B = \begin{bmatrix} 0 & 1 & 2 \\ 3 & -1 & -1 \end{bmatrix}$$

$$C = \begin{bmatrix} 1 & 2 \\ 0 & 3 \end{bmatrix} \qquad D = \begin{bmatrix} 1 & -2 & 0 & 2 \\ 1 & 0 & -1 & 1 \\ 2 & 3 & -1 & 4 \end{bmatrix}$$

$$F = \begin{bmatrix} 1 & 2 & 3 \\ 2 & -1 & 0 \\ 0 & 1 & 4 \end{bmatrix}$$

$$G = \begin{bmatrix} 0 & 1 & 2 \\ 3 & -1 & 4 \\ 2 & -1 & 0 \end{bmatrix} \qquad H = \begin{bmatrix} 1 & 2 \\ 3 & 4 \\ 5 & 6 \end{bmatrix}$$

Find

1.1 $(A + B)G$

1.2 BFG

1.3 $4A - 3B$

1.4 $AH + 3B$

1.5 AB

1.6 $A^T B$

1.7 $B^T A$

1.8 $F + G$

1.9 $(F + G)(F - G)$

1.10 $F^2 - G^2$

1.11 tr F

1.12 tr G

1.13 tr $(F + G)$

1.14 tr (FG)

1.15 Can a conclusion be drawn based on the results of exercises 1.11 through 1.14?

1.16 AA^{T}

1.17 $A^{\mathrm{T}}A$

1.18 Can a conclusion be drawn based on the results of exercises 1.15 and 1.16?

1.19 FG^{T}

In exercises 1.20 through 1.22, use the rules of matrix algebra to obtain the indicated algebraic expression:

1.20 $(F + G)(F - G)$

1.21 $(F + G)$

1.22 $FG(F + G)(G - F)$

1.23 Verify the result of exercise 1.9 using the expansion developed in exercise 1.20.

1.24 Verify the result of exercise 1.21.

1.25 Find x if

$$[1 \quad 2 \quad x] \begin{bmatrix} 1 & 2 & 0 \\ 2 & -1 & 4 \\ 3 & 0 & -1 \end{bmatrix} \begin{bmatrix} 1 \\ -2 \\ x \end{bmatrix} = [0]$$

For exercises 1.26 and 1.27 if A and B are given by

$$A = \begin{bmatrix} \cos\theta_1 & -\sin\theta_1 \\ \sin\theta_1 & \cos\theta_1 \end{bmatrix} \qquad B = \begin{bmatrix} \cos\theta_2 & -\sin\theta_2 \\ \sin\theta_2 & \cos\theta_2 \end{bmatrix}$$

1.26 Find AB

1.27 Find BA

1.28 What conclusion can be drawn from the results of exercises 1.26 and 1.27?

1.29 What, if anything, can be deduced from the results of the matrix products AB and BA if

$$A = \begin{bmatrix} 2 & 2 & 2 \\ 8 & 1 & 2 \\ 0 & -1 & 6 \end{bmatrix} \qquad B = \begin{bmatrix} 5 & -2 & 0 \\ -2 & 3 & -1 \\ 0 & -1 & 1 \end{bmatrix}$$

Exercises 1.30 to 1.36 require an example of a particular type of matrix that must be at least 3×3:

1.30 Symmetric matrix

1.31 Skew matrix

1.32 Skew-symmetric matrix

1.33 Complex matrix

1.34 Conjugate matrix

1.35 Hermitian matrix

1.36 Skew-Hermitian matrix

1.37 Why does the skew-symmetric matrix require all elements on the principal diagonal to be zero?

1.38 Why does the skew-Hermitian matrix require all elements on the principal diagonal to be purely imaginary?

DETERMINANTS

2.1 INTRODUCTION

A square matrix of order n (an $n \times n$ matrix)

$$A = \begin{bmatrix} a_{11} & a_{12} & a_{13} & \cdots & a_{1n} \\ a_{21} & a_{22} & a_{23} & \cdots & a_{2n} \\ a_{31} & a_{32} & a_{33} & \cdots & a_{3n} \\ & & \cdots & & \\ a_{n1} & a_{n2} & a_{n3} & \cdots & a_{nn} \end{bmatrix}$$

possesses a uniquely defined scalar (a single number) that is designated as the determinant of A or merely the determinant

$$\det A = |A|$$

where the order of the determinant is, by definition, the same as the order of the matrix from which it derives. Observe that only square matrices possess determinants, that vertical lines and not brackets designate determinants, and that while det A is a number and has no elements, it is customary to represent it as an array of the elements of the matrix A

$$\det A = \begin{vmatrix} a_{11} & a_{12} & a_{13} & \cdots & a_{1n} \\ a_{21} & a_{22} & a_{23} & \cdots & a_{2n} \\ a_{31} & a_{32} & a_{33} & \cdots & a_{3n} \\ & & \cdots & & \\ a_{n1} & a_{n2} & a_{n3} & \cdots & a_{nn} \end{vmatrix}$$

25

A determinant of the first order consists of a single number a and has, therefore, the value det $A = a$. A determinant of the second order is a number that derives from a 2×2 square array with the value

$$\det A = \det \begin{bmatrix} a_{11} & a_{12} \\ a_{21} & a_{22} \end{bmatrix} = \begin{vmatrix} a_{11} & a_{12} \\ a_{21} & a_{22} \end{vmatrix} = a_{11}a_{22} - a_{12}a_{21}$$

A determinant of the third order is a single number that derives from a 3×3 square array containing nine elements

$$\det A = \det \begin{bmatrix} a_{11} & a_{12} & a_{13} \\ a_{21} & a_{22} & a_{23} \\ a_{31} & a_{32} & a_{33} \end{bmatrix} = \begin{vmatrix} a_{11} & a_{12} & a_{13} \\ a_{21} & a_{22} & a_{23} \\ a_{31} & a_{32} & a_{33} \end{vmatrix}$$

One may deduce that a determinant of the nth order is a number that derives from a square array of $n \times n$ elements, a_{ij}, and that the total number of elements in an nth order determinant is n^2.

Although the concept of the determinant looks to be purely abstract, the determinant can be proved to be a very rational function that can be evaluated in a number of ways. Moreover, the value of the use of determinants in the taking of matrix inverses and in the solution of simultaneous, linear algebraic equations cannot and should not be underemphasized.

2.2 SOME EXAMPLES OF THE SOLUTION OF SIMULTANEOUS EQUATIONS

Consider, for example, the following pair of simultaneous linear algebraic equations

$$a_{11}x_1 + a_{12}x_2 = b_1 \tag{2.1a}$$

$$a_{21}x_1 + a_{22}x_2 = b_2 \tag{2.1b}$$

and observe that they may also be written in the matrix form $AX = B$

$$\begin{bmatrix} a_{11} & a_{12} \\ a_{21} & a_{22} \end{bmatrix} \begin{bmatrix} x_1 \\ x_2 \end{bmatrix} = \begin{bmatrix} b_1 \\ b_2 \end{bmatrix}$$

In Eqs. (2.1) the x's are the unknowns, and the a's form the coefficient matrix A. If det $A \neq 0$, the equations are said to be linearly independent and one method of solving this second-order system is to multiply Eq. (2.1a) by a_{22} and Eq. (2.1b) by a_{12}

$$a_{22}a_{11}x_1 + a_{22}a_{12}x_2 = a_{22}b_1$$

$$a_{12}a_{21}x_1 + a_{12}a_{22}x_2 = a_{12}b_2$$

A subtraction then yields

$$(a_{22}a_{11} - a_{12}a_{21})x_1 = a_{22}b_1 - a_{12}b_2$$

and then x_1 is obtained:

$$x_1 = \frac{a_{22}b_1 - a_{12}b_2}{a_{22}a_{11} - a_{12}a_{21}} \tag{2.2a}$$

A similar procedure yields x_2:

$$x_2 = \frac{a_{11}b_2 - a_{21}b_1}{a_{22}a_{11} - a_{12}a_{21}} \tag{2.2b}$$

Observe that the denominators of the equations that yield x_1 and x_2 can be represented by the determinant

$$\begin{vmatrix} a_{11} & a_{12} \\ a_{21} & a_{22} \end{vmatrix} = a_{11}a_{22} - a_{12}a_{21}$$

and it is easy to see that the numerators of these equations can be represented by

$$\begin{vmatrix} b_1 & a_{12} \\ b_2 & a_{22} \end{vmatrix} = b_1a_{22} - b_2a_{12}$$

and

$$\begin{vmatrix} a_{11} & b_1 \\ a_{21} & b_2 \end{vmatrix} = a_{11}b_2 - a_{21}b_1$$

This must always hold unless the determinant in the denominators is equal to zero, which is ruled out because the discussion began originally with the statement that the two equations to be solved were linearly independent ($\det A \neq 0$)

Thus one may write the solutions for x_1 and x_2 in Eqs. (2.1)

$$x_1 = \frac{\begin{vmatrix} b_1 & a_{12} \\ b_2 & a_{22} \end{vmatrix}}{\begin{vmatrix} a_{11} & a_{12} \\ a_{21} & a_{22} \end{vmatrix}} \tag{2.3a}$$

and

$$x_2 = \frac{\begin{vmatrix} a_{11} & b_1 \\ a_{21} & b_2 \end{vmatrix}}{\begin{vmatrix} a_{11} & a_{12} \\ a_{21} & a_{22} \end{vmatrix}} \tag{2.3b}$$

and this is a demonstration of what is observed before this chapter concludes of a method of solution of simultaneous linear algebraic equations known as Cramer's rule.

The foregoing reasoning applies equally well to a set of n simultaneous algebraic equations. Only the difficulty is at stake. For a set of three equations in three unknowns

$$a_{11}x_1 + a_{12}x_2 + a_{13}x_3 = b_1$$

$$a_{21}x_1 + a_{22}x_2 + a_{23}x_3 = b_2$$

$$a_{31}x_1 + a_{32}x_2 + a_{33}x_3 = b_3$$

that are assumed to be linearly independent and that may be written in matrix form

$$\begin{bmatrix} a_{11} & a_{12} & a_{13} \\ a_{21} & a_{22} & a_{23} \\ a_{31} & a_{32} & a_{33} \end{bmatrix} \begin{bmatrix} x_1 \\ x_2 \\ x_3 \end{bmatrix} = \begin{bmatrix} b_1 \\ b_2 \\ b_3 \end{bmatrix}$$

It can be shown that x_1 can be evaluated from

$$x_1 = \frac{b_1a_{22}a_{33} + b_3a_{12}a_{23} + b_2a_{13}a_{32} - b_3a_{22}a_{13} - b_1a_{32}a_{23} - b_2a_{12}a_{33}}{a_{11}a_{22}a_{33} + a_{12}a_{23}a_{31} + a_{13}a_{21}a_{32} - a_{31}a_{22}a_{13} - a_{32}a_{23}a_{11} - a_{33}a_{21}a_{12}}$$

Both the numerator and denominator can be rearranged by employing a little algebra

$$x_1 = \frac{b_1(a_{22}a_{33} - a_{32}a_{23}) - b_2(a_{12}a_{33} - a_{32}a_{13}) + b_3(a_{12}a_{23} - a_{22}a_{13})}{a_{11}(a_{22}a_{33} - a_{32}a_{23}) - a_{21}(a_{12}a_{33} - a_{32}a_{13}) + a_{31}(a_{12}a_{23} - a_{22}a_{13})}$$

and an inspection of the terms within the parentheses shows that the solution for x_1 can not only be written (Cramer's rule) as the quotient of two determinants but each of the determinants can be represented in terms of three second-order determinants

$$x_1 = \frac{\begin{vmatrix} b_1 & a_{12} & a_{13} \\ b_2 & a_{22} & a_{23} \\ b_3 & a_{32} & a_{33} \end{vmatrix} = b_1\begin{vmatrix} a_{22} & a_{23} \\ a_{32} & a_{33} \end{vmatrix} - b_2\begin{vmatrix} a_{12} & a_{13} \\ a_{32} & a_{33} \end{vmatrix} + b_3\begin{vmatrix} a_{12} & a_{13} \\ a_{22} & a_{23} \end{vmatrix}}{\begin{vmatrix} a_{11} & a_{12} & a_{13} \\ a_{21} & a_{22} & a_{23} \\ a_{31} & a_{32} & a_{33} \end{vmatrix} = a_{11}\begin{vmatrix} a_{22} & a_{23} \\ a_{32} & a_{33} \end{vmatrix} - a_{21}\begin{vmatrix} a_{12} & a_{13} \\ a_{32} & a_{33} \end{vmatrix} + a_{31}\begin{vmatrix} a_{12} & a_{13} \\ a_{22} & a_{23} \end{vmatrix}}$$

$$(2.4)$$

This expansion is known as the Laplace expansion or the Laplace development, which is discussed in detail as this study of determinants proceeds.

2.3 EVALUATION OF SECOND- AND THIRD-ORDER DETERMINANTS

The method of evaluating second-order determinants is suggested in Eqs. (2.2) and (2.4). The second-order determinant is evaluated as the remainder of the product resulting from the multiplication of the upper-left and lower-right elements (the principal diagonal elements) minus the product of the lower-left and the upper-right elements (the secondary diagonal elements). This procedure is demonstrated in Fig. 2.1a.

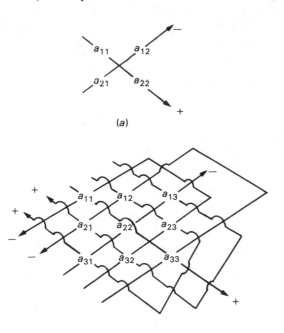

(a)

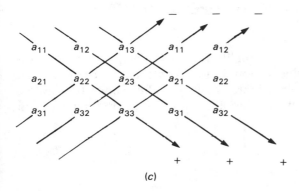

(b)

(c)

Figure 2.1 (a) Procedure for evaluating a second-order determinant; (b) or (c) equivalent procedures for evaluating a third-order determinant.

The third-order determinant may be evaluated by taking the products and then the sums and differences of the elements shown in Fig. 2.1b. This procedure may be assisted by rewriting the first two columns of the determinant and then proceeding as indicated in Fig. 2.1c. It is important to note that for this purpose the diagonals of the third-order determinant are continuous, that is, the last column is followed by the first column.

Caution is necessary: fourth- and higher-order determinants may not be evaluated by following the procedures displayed in Fig. 2.1.

Example: The second-order determinant

$$|A| = \begin{vmatrix} 2 & 3 \\ 1 & 4 \end{vmatrix}$$

is evaluated by the procedure indicated in Fig. 2.1a

$$\begin{vmatrix} 2 & 3 \\ 1 & 4 \end{vmatrix} = 2(4) - 3(1) = 8 - 3 = 5$$

The third-order determinant

$$|B| = \begin{vmatrix} 3 & 2 & 1 \\ 1 & 4 & 2 \\ 1 & 3 & 1 \end{vmatrix}$$

is evaluated in accordance with Figs. 2.1b or c as

$$|B| = 3(4)(1) + 2(2)(1) + 1(1)(3) - 1(4)(1) - 3(2)(3) - 1(1)(2)$$

or

$$|B| = 12 + 4 + 3 - 4 - 18 - 2 = -5$$

2.4 PRELIMINARY RULES FOR OPERATION WITH DETERMINANTS

Several rules pertaining to the simplification and manipulation of determinants are presented in this section without formal proof. Their validity, however, is amply demonstrated through the use of examples that involve the third-order determinant used in the example in the preceding section

$$B = \begin{vmatrix} 3 & 2 & 1 \\ 1 & 4 & 2 \\ 1 & 3 & 1 \end{vmatrix}$$

Rule One: Interchanging any row (or column) of a determinant with its immediately adjacent row (or column) alters the sign of the determinant.

Example: If the second and third rows of $|B|$ are interchanged, then

$$\begin{vmatrix} 3 & 2 & 1 \\ 1 & 3 & 1 \\ 1 & 4 & 2 \end{vmatrix} = 18 + 2 + 4 - 3 - 12 - 4 = 24 - 19 = 5$$

Rule Two: The multiplication of any single row (column) of a determinant by a scalar constant is equivalent to the multiplication of the entire determinant by the scalar.

Observe that this differs from the multiplication of a matrix by a scalar; the multiplication of a matrix by a scalar results in the multiplication of each and every element of the matrix by the scalar.

Example: Suppose that the first column of $|B|$ is multiplied by 2. Then

$$\begin{vmatrix} 2(3) & 2 & 1 \\ 2(1) & 4 & 2 \\ 2(1) & 3 & 1 \end{vmatrix} = 24 + 8 + 6 - 8 - 36 - 4 = 38 - 48 = -10$$

This, of course, is equal to

$$(2) \begin{vmatrix} 3 & 2 & 1 \\ 1 & 4 & 2 \\ 1 & 3 & 1 \end{vmatrix} = 2(-5) = -10$$

If every element in an nth order determinant is multiplied by the same scalar, α, the value of the determinant is multiplied by α^n.

Example: If every element of $|B|$ is multiplied by 2:

$$\begin{vmatrix} 2(3) & 2(2) & 2(1) \\ 2(1) & 2(4) & 2(2) \\ 2(1) & 2(3) & 2(1) \end{vmatrix} = \begin{vmatrix} 6 & 4 & 2 \\ 2 & 8 & 4 \\ 2 & 6 & 2 \end{vmatrix} = 96 + 32 + 24 - 32 - 144 - 16$$

or $152 - 192 = -40$, which is clearly $(2)^3(-5) = 8(-5) = -40$.

Rule Three: If any two rows (columns) of a determinant are identical, the value of the determinant is zero and the matrix from which the determinant derives is said to be singular.

Example: Notice that in

$$\begin{vmatrix} 3 & 3 & 1 \\ 1 & 1 & 2 \\ 1 & 1 & 1 \end{vmatrix} = 3 + 6 + 1 - 1 - 6 - 3 = 0 - 0 = 0$$

the first and second columns are identical and the value of the determinant is zero.

Rule Four: If any row (or column) of a determinant contains nothing but zeroes, the value of the determinant is zero.

Example: The determinant that has a column of zeroes in its center is seen to have a value equal to zero:

$$\begin{vmatrix} 3 & 0 & 1 \\ 1 & 0 & 2 \\ 4 & 0 & -3 \end{vmatrix} = 0 + 0 + 0 - 0 - 0 - 0 = 0$$

Rule Five: If any two rows (columns) of a determinant are proportional,† the determinant is equal to zero.

Example: In the determinant

$$|D| = \begin{vmatrix} 3 & 9 & 1 \\ 1 & 3 & 2 \\ 2 & 6 & -1 \end{vmatrix}$$

notice that the values of the elements in column two are exactly three times the value of the corresponding elements in the first column. The value of this determinant is

$$|D| = -9 + 36 + 6 - 6 - 36 + 9 = 0 - 0 = 0$$

and the matrix from which it derives is singular.

Rule Six: If the elements of any row (column) of a determinant are added to or subtracted from the corresponding elements of another row (column), the value of the determinant is unchanged.

Example: In the determinant

$$|B| = \begin{vmatrix} 3 & 2 & 1 \\ 1 & 4 & 2 \\ 1 & 3 & 1 \end{vmatrix}$$

†The two rows (columns) are said to be linearly dependent. This important topic is addressed in Chapter 5.

add the elements of the third row to the first row. This creates a new determinant:

$$|C| = \begin{vmatrix} (3+1) & (2+3) & (1+1) \\ 1 & 4 & 2 \\ 1 & 3 & 1 \end{vmatrix} = \begin{vmatrix} 4 & 5 & 2 \\ 1 & 4 & 2 \\ 1 & 3 & 1 \end{vmatrix}$$

but its value

$$|C| = 16 + 10 + 6 - 8 - 24 - 5 = 32 - 37 = -5$$

is unchanged.

Rules two and six combine to produce the same effect.

Rule Six(a): If the elements of any row (column) of a determinant are multiplied by a constant and then added to or subtracted from the corresponding elements of another row (column), the value of the determinant is unchanged.

Example: This time in

$$|B| = \begin{vmatrix} 3 & 2 & 1 \\ 1 & 4 & 2 \\ 1 & 3 & 1 \end{vmatrix}$$

multiply column two by 3 and add the result to column three. The result is

$$|C| = \begin{vmatrix} 3 & 2 & (3 \times 2 + 1) \\ 1 & 4 & (3 \times 4 + 2) \\ 1 & 3 & (3 \times 3 + 1) \end{vmatrix} = \begin{vmatrix} 3 & 2 & 7 \\ 1 & 4 & 14 \\ 1 & 3 & 10 \end{vmatrix}$$

The value of this new determinant

$$|C| = 120 + 28 + 21 - 28 - 126 - 20 = 169 - 174 = -5$$

is unchanged.

Rule Seven: The value of the determinant of a diagonal matrix is equal to the product of the terms on the diagonal.

Example: In

$$\begin{vmatrix} 2 & 0 & 0 \\ 0 & -1 & 0 \\ 0 & 0 & 9 \end{vmatrix}$$

it is easy to see that the value of the determinant is -18.

Rule Eight: The value of the determinant of a matrix is equal to the value of the determinant of the transpose of the matrix.

Example: Consider the matrix B and its transpose

$$B = \begin{bmatrix} 3 & 2 & 1 \\ 1 & 4 & 2 \\ 1 & 3 & 1 \end{bmatrix} \qquad B^\mathrm{T} = \begin{bmatrix} 3 & 1 & 1 \\ 2 & 4 & 3 \\ 1 & 2 & 1 \end{bmatrix}$$

The determinant of the transpose is seen to be

$$\det (B^\mathrm{T}) = 12 + 3 + 4 - 4 - 18 - 2 = 19 - 24 = -5$$

which agrees with $\det B = -5$.

Rule Nine: The determinant of the product of two matrices is equal to the product of the determinants of the two matrices.

Example: Consider two matrices A and C:

$$A = \begin{bmatrix} 2 & 3 & -1 \\ 1 & 2 & 3 \\ 0 & -4 & 5 \end{bmatrix} \qquad B^\mathrm{T} = C = \begin{bmatrix} 3 & 1 & 1 \\ 2 & 4 & 3 \\ 1 & 2 & 1 \end{bmatrix}$$

The value of $\det C$ is known:

$$\det C = -5$$

and the value of $\det A$ is

$$\det A = 20 + 0 + 4 - 0 + 24 - 15 = 33$$

According to rule nine the value of the determinant of the product AC is

$$\det AC = (\det A)(\det C) = 33(-5) = -165$$

and this can be verified by taking AC and then CA:

$$AC = \begin{bmatrix} 2 & 3 & -1 \\ 1 & 2 & 3 \\ 0 & -4 & 5 \end{bmatrix}\begin{bmatrix} 3 & 1 & 1 \\ 2 & 4 & 3 \\ 1 & 2 & 1 \end{bmatrix} = \begin{bmatrix} 11 & 12 & 10 \\ 10 & 15 & 10 \\ -3 & -6 & -7 \end{bmatrix}$$

and

$$\det AC = -1115 - 360 - 600 + 450 + 660 + 840$$
$$= 1950 - 2115 = -165$$

Now get CA:

$$CA = \begin{bmatrix} 3 & 1 & 1 \\ 2 & 4 & 3 \\ 1 & 2 & 1 \end{bmatrix} \begin{bmatrix} 2 & 3 & -1 \\ 1 & 2 & 3 \\ 0 & -4 & 5 \end{bmatrix} = \begin{bmatrix} 7 & 7 & 5 \\ 8 & 2 & 25 \\ 4 & 3 & 10 \end{bmatrix}$$

and then det CA

$$\det CA = 140 + 700 + 120 - 40 - 525 - 560$$
$$= 960 - 1125 = -165$$

This demonstrates that although, in general, matrix multiplication is not commutative, determinant multiplication is commutative; that is det AC = det CA = (det A)(det C).

However, in general, det $(A + C) \neq$ det A + det C:

$$A + C = \begin{bmatrix} 2 & 3 & -1 \\ 1 & 2 & 3 \\ 0 & -4 & 5 \end{bmatrix} + \begin{bmatrix} 3 & 1 & 1 \\ 2 & 4 & 3 \\ 1 & 2 & 1 \end{bmatrix} = \begin{bmatrix} 5 & 4 & 0 \\ 3 & 6 & 6 \\ 1 & -2 & 6 \end{bmatrix}$$

and

$$\det (A + C) = 180 + 24 + 0 - 0 + 60 - 72 = 264 - 72 = 192$$

where

$$\det A + \det C = 33 - 5 = 28$$

Rule Ten: If the determinant of the product of two square matrices is zero, then at least one of the matrices is singular, that is, the value of its determinant is equal to zero.

Example: Recall that in the discussion of rule five, it was shown that

$$\det D = \det \begin{bmatrix} 3 & 9 & 1 \\ 1 & 3 & 2 \\ 2 & 6 & 1 \end{bmatrix} = 0$$

Take the product DC and obtain

$$\begin{bmatrix} 3 & 9 & 1 \\ 1 & 3 & 2 \\ 2 & 6 & 1 \end{bmatrix}\begin{bmatrix} 3 & 1 & 1 \\ 2 & 4 & 3 \\ 1 & 2 & 1 \end{bmatrix} = \begin{bmatrix} 28 & 41 & 31 \\ 11 & 17 & 12 \\ 19 & 28 & 21 \end{bmatrix}$$

Then

$$\det DB = 9996 + 9348 + 9548 - 10{,}013 - 9408 - 9471$$
$$= 28{,}892 - 28{,}892 = 0$$

Rule Eleven: If an $m \times n$ rectangular matrix A is postmultiplied by an $n \times m$ rectangular matrix B, the resulting square matrix $C = AB$ of order m will, in general, be singular if $m > n$.

Example: This time take two entirely different matrices, A which is 3 $\times$ 2 and B which is 2 $\times$ 3

$$A = \begin{bmatrix} 1 & 2 \\ 3 & 4 \\ -1 & -1 \end{bmatrix} \qquad B = \begin{bmatrix} 1 & 2 & 0 \\ -1 & -1 & 3 \end{bmatrix}$$

and obtain the product $C = AB$ which will be 3 $\times$ 3, a square matrix of third order

$$C = AB = \begin{bmatrix} 1 & 2 \\ 3 & 4 \\ -1 & -1 \end{bmatrix}\begin{bmatrix} 1 & 2 & 0 \\ -1 & -1 & 3 \end{bmatrix}\begin{bmatrix} -1 & 0 & 6 \\ -1 & 2 & 12 \\ 0 & -1 & -3 \end{bmatrix}$$

Here $m > n$ and

$$\det C = 6 + 0 + 6 - 0 - 12 - 0 = 12 - 12 = 0$$

which is an instance of rule eleven.

The product $D = BA$ which is 2 $\times$ 2 and which involves a matrix multiplication where $m < n$ is, however,

$$D = BA = \begin{bmatrix} 1 & 2 & 0 \\ -1 & -1 & 3 \end{bmatrix}\begin{bmatrix} 1 & 2 \\ 3 & 4 \\ -1 & -1 \end{bmatrix} = \begin{bmatrix} 7 & 10 \\ -7 & -9 \end{bmatrix}$$

and possesses a determinant

$$\det D = -63 + 70 = 7$$

that is not zero.

2.5 MINORS, COMPLEMENTARY MINORS, ALGEBRAIC COMPLEMENTS, AND COFACTORS

Consider the nth-order determinant

$$|A| = \begin{vmatrix} a_{11} & a_{12} & a_{13} & a_{14} & \cdots & a_{1n} \\ a_{21} & a_{22} & a_{23} & a_{24} & \cdots & a_{2n} \\ a_{31} & a_{32} & a_{33} & a_{34} & \cdots & a_{3n} \\ a_{41} & a_{42} & a_{43} & a_{44} & \cdots & a_{4n} \\ & & \cdots & & \\ a_{n1} & a_{n2} & a_{n3} & a_{n4} & \cdots & a_{nn} \end{vmatrix} \tag{2.5}$$

which is used for proposing several useful quantities.

The mth-order minor of an nth-order matrix is the determinant formed by deleting $n - m$ rows and $n - m$ columns in the nth-order determinant. For example, the minor $|M|_{ir}$ of the determinant $|A|$ is formed by deleting the ith row and the rth column of $|A|$. Because $|A|$ is of order n, the minor $|M|_{ir}$ is of order $m = n - 1$ and contains m^2 elements. In general terms, a minor formed by deleting p rows and p columns in the nth-order determinant $|A|$ is called an $(n - p)$th-order minor. If $p = n - 1$, the minor is of first order and consists of a single element of $|A|$. From this it is easy to see that the nth-order determinant $|A|$ contains n^2 first-order minors, each consisting of a single element.

When dealing with minors other than those of $(n - 1)$th order, the designation of the eliminated rows and columns of the determinant $|A|$ must be considered very carefully, or the results can be chaotic and lead to disappointment. It is best to consider rows $j, k, l, m, \ldots$ and columns $r, s, t, u, \ldots$ so that the $(n - 1)$th-, the $(n - 2)$th-, and the $(n - 3)$th-order minors would be designated, respectively, as $|M|_{j,r}$, $|M|_{jk,rs}$, and $|M|_{jkl,rst}$. Particular note should be taken of the subscript notation.

Example: In the nth-order determinant given by Eq. (2.5), the minor formed by deleting the second row and the third column is

$$|M|_{23} = \begin{vmatrix} a_{11} & a_{12} & a_{14} & \cdots & a_{1n} \\ a_{31} & a_{32} & a_{34} & \cdots & a_{3n} \\ a_{41} & a_{42} & a_{44} & \cdots & a_{4n} \\ & & \cdots & & \\ a_{n1} & a_{n2} & a_{n4} & \cdots & a_{nn} \end{vmatrix} \tag{2.6}$$

This minor is of $(n - 1)$th order and possesses a total of $(n - 1)^2$ elements.

If, in addition, the third row and first column of $|A|$ are deleted, an $(n - 2)$th-order minor containing $(n - 2)^2$ elements can be formed:

$$|M|_{23,31} = \begin{vmatrix} a_{12} & a_{14} & \cdots & a_{1n} \\ a_{42} & a_{44} & \cdots & a_{4n} \\ & & \cdots & \\ a_{n2} & a_{n4} & \cdots & a_{nn} \end{vmatrix} \tag{2.7}$$

Particular attention should be paid to the use of the subscripts.

The complementary minor or complement of a minor, designated as $|N|$ (with subscript) is the determinant formed by placing the elements that lie at the intersections of the deleted rows and columns of the original determinant into a square array in the order in which these elements appear in the original determinant.

Example: For the nth-order determinant $|A|$ shown in Eq. (2.5), two minors were formed in the previous example. The $(n - 1)$th-order minor given by Eq. (2.6) has a complementary minor or complement of first order consisting of a single element

$$|N|_{23} = a_{23}$$

The $(n - 2)$th-order minor displayed in Eq. (2.7) has a second-order complementary minor or complement

$$|N|_{23,31} = \begin{vmatrix} a_{21} & a_{23} \\ a_{31} & a_{33} \end{vmatrix}$$

The algebraic complement of the minor $|M|$ is the "signed" complementary minor. If a minor is formed from the nth-order determinant $|A|$ by deleting rows i, k, l, . . . and columns r, s, t, . . . from the determinant $|A|$, the minor is designated as

$$|M|_{ikl,rst}$$

the complementary minor or complement is designated as

$$|N|_{ikl,\text{rst}}$$

and the algebraic complement is defined as

$$(-1)^{i+k+l+\cdots+r+s+t}|N|_{ikl,rst}$$

The cofactor, designated as A, without vertical rules and with a double subscript, is the "signed" $(n - 1)$th minor formed from an nth-order determinant. Suppose that the $(n - 1)$th-order minor has been formed by deleting the ith row and the jth column from the determinant $|A|$. Then

$$A_{ij} = (-1)^{i+j}|M|_{ij} \tag{2.8}$$

Observe that the cofactor has no meaning for minors of order smaller than $(n - 1)$ unless the minor itself is being treated as a determinant of order one less than the determinant $|A|$ from which it derives. Also observe that when the minor is of order $(n - 1)$, the product of the cofactor and the complement is equal to the product of the minor and the algebraic complement.

Example: The fourth-order determinant

$$|A| = \begin{vmatrix} 1 & 3 & -1 & 2 \\ 4 & 1 & 1 & 3 \\ 3 & 1 & -2 & 1 \\ 1 & 3 & 2 & 5 \end{vmatrix}$$

has a third-order minor formed by deleting the third row and the fourth column:

$$|M|_{34} = \begin{vmatrix} 1 & 3 & -1 \\ 4 & 1 & 1 \\ 1 & 3 & 2 \end{vmatrix} = 2 + 3 - 12 + 1 - 3 - 24 = 6 - 39 = -33$$

which in turn has a first-order complementary minor

$$|N|_{34} = 1$$

The algebraic complement of the minor $|M|_{34}$ is

$$(-1)^{3+4}(1) = (-1)^7(1) = -1$$

and the cofactor is the signed minor

$$A_{34} = (-1)^{3+4}(-33) = -(-33) = 33$$

If, in addition, the first row and the first column are deleted from $|A|$, a second-order minor

$$|M|_{31,41} = \begin{vmatrix} 1 & 1 \\ 3 & 2 \end{vmatrix} = 2 - 3 = -1$$

is obtained, as is a second-order complementary minor

$$|N|_{31,41} = \begin{vmatrix} 1 & 2 \\ 3 & 1 \end{vmatrix} = 1 - 6 = -5$$

In this case the algebraic complement is

$$(-1)^{3+1+4+1}(-5) = -(-5) = 5$$

but a cofactor, in this case, does not exist.

In the forgoing example, it should be noted that when dealing with the third-order minor, $|M|_{34} = -33$, the product of the minor and its algebraic complement, $(-33)(-1) = 33$, is equal to the product of the cofactor and the complementary minor, $(1)(33) = 33$.

2.6 THE COFACTOR MATRIX

A square nth-order matrix such as the one displayed in Eq. (2.1) possesses a cofactor matrix with elements indicated by capital letters with double subscripts

$$A^C = \begin{bmatrix} A_{11} & A_{12} & A_{13} & A_{14} & \cdots & A_{1n} \\ A_{21} & A_{22} & A_{23} & A_{24} & \cdots & A_{2n} \\ A_{31} & A_{32} & A_{33} & A_{34} & \cdots & A_{3n} \\ A_{41} & A_{42} & A_{43} & A_{44} & \cdots & A_{4n} \\ & & \cdots & & & \\ A_{n1} & A_{n2} & A_{n3} & A_{n4} & \cdots & A_{nn} \end{bmatrix} \qquad (2.9)$$

Example: The symmetrical matrix

$$A = \begin{bmatrix} 3 & -2 & 0 \\ -2 & 4 & -1 \\ 0 & -1 & 6 \end{bmatrix}$$

has a determinant

$$\det A = 72 + 0 + 0 - 0 - 3 - 24 = 72 - 27 = 45$$

and nine cofactors whose sign can be determined either from Eq. (2.8) or from the "checkerboard" shown in Fig. 2.2

$$A_{11} = +|M|_{11} = + \begin{vmatrix} 4 & -1 \\ -1 & 6 \end{vmatrix} = +(24 - 1) = 23$$

$$A_{12} = -|M|_{12} = - \begin{vmatrix} -2 & -1 \\ 0 & 6 \end{vmatrix} = -(-12) = 12$$

$$A_{13} = +|M|_{13} = + \begin{vmatrix} -2 & 4 \\ 0 & -1 \end{vmatrix} = 2$$

$$A_{21} = -|M|_{21} = - \begin{vmatrix} -2 & 0 \\ -1 & 6 \end{vmatrix} = -(-12) = 12$$

$$A_{22} = +|M|_{22} = + \begin{vmatrix} 3 & 0 \\ 0 & 6 \end{vmatrix} = 18$$

$$A_{23} = -|M|_{23} = - \begin{vmatrix} 3 & -2 \\ 0 & -1 \end{vmatrix} = -(-3) = 3$$

$$A_{31} = +|M|_{31} = + \begin{vmatrix} -2 & 0 \\ 4 & -1 \end{vmatrix} = 2$$

$$A_{32} = -|M|_{32} = - \begin{vmatrix} 3 & 0 \\ -2 & -1 \end{vmatrix} = -(-3) = 3$$

$$A_{33} = +|M|_{33} = + \begin{vmatrix} 3 & -2 \\ -2 & 4 \end{vmatrix} = +(12 - 4) = 8$$

Thus

$$A^C = \begin{bmatrix} 23 & 12 & 2 \\ 12 & 18 & 3 \\ 2 & 3 & 8 \end{bmatrix}$$

and it can be observed that symmetrical matrices possess symmetrical co-factor matrices.

Figure 2.2 Checkerboard rule for finding the sign of a cofactor of an nth-order determinant, (a) for n odd and (b) for n even.

2.7 RULES FOR OPERATIONS WITH COFACTORS

In the denominator of Eq. (2.4) the third-order determinant $|A|$ was shown to be equal to a certain function of three second-order determinants

$$|A| = \begin{vmatrix} a_{11} & a_{12} & a_{13} \\ a_{21} & a_{22} & a_{23} \\ a_{31} & a_{32} & a_{33} \end{vmatrix} = a_{11}\begin{vmatrix} a_{22} & a_{23} \\ a_{32} & a_{33} \end{vmatrix} - a_{21}\begin{vmatrix} a_{12} & a_{13} \\ a_{32} & a_{33} \end{vmatrix} + a_{31}\begin{vmatrix} a_{12} & a_{13} \\ a_{22} & a_{23} \end{vmatrix}$$

Notice that each of the second-order determinants is a second-order minor of $|A|$. This means that three cofactors exist as defined by Eq. (2.8) and, hence, for the third-order determinant

$$|A| = \begin{vmatrix} a_{11} & a_{12} & a_{13} \\ a_{21} & a_{22} & a_{23} \\ a_{31} & a_{32} & a_{33} \end{vmatrix} = a_{11}|M|_{11} - a_{21}|M|_{21} + a_{31}|M|_{31}$$

or

$$|A| = a_{11}A_{11} + a_{21}A_{21} + a_{31}A_{31}$$

Rule Twelve: A determinant may be evaluated by summing the products of every element in any row (column) by its respective cofactor.†

Example: The fourth-order determinant

$$|A| = \begin{vmatrix} 6 & 3 & 0 & 1 \\ 2 & 5 & 1 & -2 \\ 1 & 0 & -1 & 1 \\ 1 & 0 & 3 & 4 \end{vmatrix}$$

can be expanded using the second column to reduce the labor (two zeroes occur in this column):

$$|A| = a_{21}A_{21} + a_{22}A_{22}$$

The cofactors derive from the appropriate minors with their sign determined from Eq. (2.8) or from the checkerboard rule illustrated in Fig. 2.2.

$$A_{21} = -|M|_{21} = -\begin{vmatrix} 2 & 1 & -2 \\ 1 & -1 & 1 \\ 1 & 3 & 4 \end{vmatrix} = -(-8 + 1 - 6 - 2 - 6 - 4)$$

†This is the procedure known as the Laplace development or the Laplace expansion.

or

$$A_{21} = -(1 - 26) = -(-25) = 25$$

and

$$A_{22} = +|M|_{22} = + \begin{vmatrix} 6 & 0 & 1 \\ 1 & -1 & 1 \\ 1 & 3 & 4 \end{vmatrix} = (-24 + 0 + 3 + 1 - 18 - 0)$$

or

$$A_{22} = 4 - 42 = -38$$

The value of the determinant is

$$|A| = a_{21}A_{21} + a_{22}A_{22}$$

$$= 3(25) + 5(-38) = 75 - 190 = -115$$

Rule Thirteen: If all cofactors in a row (column) are equal to zero, the determinant is equal to zero and the matrix from which it derives is singular.

Rule Fourteen: If the elements of a row (column) of a determinant are multiplied by cofactors of the corresponding elements of a different row (column), the resulting sum of these products is zero.

Example: In the previous example, two cofactors in the second column were identified:

$$A_{21} = 25$$

and

$$A_{22} = -38$$

Two other cofactors may be obtained:

$$A_{23} = - \begin{vmatrix} 6 & 0 & 1 \\ 2 & 1 & -2 \\ 1 & 3 & 4 \end{vmatrix} = -(24 + 0 + 6 - 1 + 36 - 0)$$

or

$$A_{23} = -(66 - 1) = -65$$

and

$$A_{24} = + \begin{vmatrix} 6 & 0 & 1 \\ 2 & 1 & -2 \\ 1 & -1 & 1 \end{vmatrix} = (6 + 0 - 2 - 1 - 12 - 0)$$

or

$$A_{24} = 6 - 15 = -9$$

The product of these cofactors with the elements of the first column leads to a zero sum:

$$6(25) + 2(-38) + 1(-65) + 1(-9) = 150 - 150 = 0$$

and a similar calculation involving the elements of the third and fourth columns leads to the same result:

$$0(25) + 1(-38) - 1(-65) + 3(-9) = 65 - 65 = 0$$

$$1(25) - 2(-38) + 1(-65) + 4(-9) = 101 - 101 = 0$$

2.8 DETERMINANT EVALUATION BY PIVOTAL CONDENSATION

The evaluation of a determinant by the Laplace expansion (Rule Twelve) can be a long, tedious, laborious procedure. Assuming that third-order determinants can be evaluated quickly, a fifth-order determinant containing no zero elements requires the evaluation of $5 \times 4 = 20$ third-order determinants. For a sixth-order determinant, this number becomes $6 \times 5 \times 4 = 120$. In general, the evaluation of an nth-order determinant can require the evaluation of $(n - 1)!$ third-order determinants, as long as $n > 6$.

Pivotal condensation is a much more efficient process and is based on Rules Two and Six in Section 2.4. Take the determinant

$$|A| = \begin{vmatrix} a_{11} & a_{12} & a_{13} & \cdots & a_{1n} \\ a_{21} & a_{22} & a_{23} & \cdots & a_{2n} \\ a_{31} & a_{32} & a_{33} & \cdots & a_{3n} \\ & & \cdots & & \\ a_{n1} & a_{n2} & a_{n3} & \cdots & a_{nn} \end{vmatrix}$$

The element a_{11} is selected as the element in the pivotal position. It is called the *pivotal element* or merely the *pivot* in the following development. The objective is to find a determinant $|B|$ that is one order less than $|A|$ by operating on $|A|$ in such a manner as to produce a column of zeroes in the column containing the pivot.

The first step is to multiply every element in every row but the first by a_{11}. In accordance with Rule Two,

$$|A| = \frac{1}{a_{11}^{n-1}} \begin{vmatrix} a_{11} & a_{12} & a_{13} & \cdots & a_{1n} \\ a_{11}a_{21} & a_{11}a_{22} & a_{11}a_{23} & \cdots & a_{11}a_{2n} \\ a_{11}a_{31} & a_{11}a_{32} & a_{11}a_{33} & \cdots & a_{11}a_{3n} \\ & & \cdots & & \\ a_{11}a_{n1} & a_{11}a_{n2} & a_{11}a_{n3} & \cdots & a_{11}a_{nn} \end{vmatrix}$$

If $a_{11} = 0$, a row or column interchange can be performed in accordance with Rule one to put a nonzero element in the pivotal position.

The next step is to multiply row one in turn by a_{i1}, where $i = 2, 3, 4, \ldots, n$ and subtract the result from each row but row one in accordance with Rule Six. The result is

$$|A| = \frac{1}{a_{11}^{n-1}}$$

$$\begin{vmatrix} a_{11} & a_{12} & a_{13} & \cdots & a_{1n} \\ 0 & (a_{11}a_{22} - a_{21}a_{12}) & (a_{11}a_{23} - a_{21}a_{13}) & \cdots & (a_{11}a_{2n} - a_{21}a_{1n}) \\ 0 & (a_{11}a_{32} - a_{31}a_{12}) & (a_{11}a_{33} - a_{31}a_{13}) & \cdots & (a_{11}a_{3n} - a_{31}a_{1n}) \\ & & \cdots & & \\ 0 & (a_{11}a_{n2} - a_{n1}a_{12}) & (a_{11}a_{n3} - a_{n1}a_{13}) & \cdots & (a_{11}a_{nn} - a_{n1}a_{1n}) \end{vmatrix}$$

and this can be expanded in accordance with Rule Twelve:

$$|A| = \frac{1}{a_{11}^{n-1}} a_{11}|M|_{11} = \frac{1}{a_{11}^{n-2}} |B|$$

where the expansion is done down the first column, in which all elements but a_{11} are equal to zero. Notice that the determinant $|B|$ in the foregoing is of order $n - 1$:

$$|B| = \begin{vmatrix} (a_{11}a_{22} - a_{21}a_{12}) & (a_{11}a_{23} - a_{21}a_{13}) & \cdots & (a_{11}a_{2n} - a_{21}a_{1n}) \\ (a_{11}a_{32} - a_{31}a_{12}) & (a_{11}a_{33} - a_{31}a_{13}) & \cdots & (a_{11}a_{3n} - a_{31}a_{1n}) \\ & \cdots & & \\ (a_{11}a_{n2} - a_{n1}a_{12}) & (a_{11}a_{n3} - a_{n1}a_{13}) & \cdots & (a_{11}a_{nn} - a_{n1}a_{1n}) \end{vmatrix}$$

Thus the condensation process brings an nth-order determinant down to an $(n - 1)$th-order determinant, and the process may be continued until the order is reduced to three or two at which point the evaluation can be completed by using the methods described in Section 2.3. The entire procedure can be handled by the computationally efficient matrix relationship:

$$|A| = \frac{1}{a^{n-2}} \det \left[a_{11} \begin{bmatrix} a_{22} & a_{23} & \cdots & a_{2n} \\ a_{32} & a_{33} & \cdots & a_{3n} \\ & & \cdots & \\ a_{n2} & a_{n3} & \cdots & a_{nn} \end{bmatrix} - \begin{bmatrix} a_{21} \\ a_{31} \\ \cdots \\ a_{n1} \end{bmatrix} \begin{bmatrix} a_{12} & a_{13} & \cdots & a_{1n} \end{bmatrix} \right] \qquad (2.10)$$

Example: Consider the determinant with a single element in its first column:

$$|A| = \begin{vmatrix} 2 & -1 & 1 & 1 & 2 \\ 0 & 2 & 3 & 2 & 1 \\ 0 & 1 & 2 & 1 & 2 \\ 0 & 1 & -1 & -1 & 3 \\ 0 & 2 & 1 & 1 & -2 \end{vmatrix}$$

This determinant is evaluated in this example by both the Laplace expansion and the method of pivotal condensation. The deliberate injection of the zeroes in all but one entry of the first column reduces the labor involved when using the Laplace expansion and still demonstrates the usefulness of pivotal condensation.

With the Laplace expansion, the first step is to reduce the fifth-order determinant to one of fourth order by expanding down the first column:

$$|A| = 2 \begin{vmatrix} 2 & 3 & 2 & 1 \\ 1 & 2 & 1 & 2 \\ 1 & -1 & -1 & 3 \\ 2 & 1 & 1 & -2 \end{vmatrix}$$

The fourth-order determinant is then reduced to four third-order determinants by expanding along the first row:

$$|A| = 2 \left\{ 2 \begin{vmatrix} 2 & 1 & 2 \\ -1 & -1 & 3 \\ 1 & 1 & -2 \end{vmatrix} - 3 \begin{vmatrix} 1 & 1 & 2 \\ 1 & -1 & 3 \\ 2 & 1 & -2 \end{vmatrix} \right.$$

$$\left. + 2 \begin{vmatrix} 1 & 2 & 2 \\ 1 & -1 & 3 \\ 2 & 1 & -2 \end{vmatrix} - \begin{vmatrix} 1 & 2 & 1 \\ 1 & -1 & -1 \\ 2 & 1 & 1 \end{vmatrix} \right\}$$

with the result

$$A = 2[2(-1) - 3(13) + 2(21) - (-3)]$$
$$= 2(-2 - 39 + 42 + 3) = 2(4) = 8$$

By pivotal condensation, the determinant

$$\det \begin{bmatrix} 2 & -1 & 1 & 1 & 2 \\ 0 & 2 & 3 & 2 & 1 \\ 0 & 1 & 2 & 1 & 2 \\ 0 & 1 & -1 & -1 & 3 \\ 0 & 2 & 1 & 1 & -2 \end{bmatrix}$$

is equal to

$$\frac{1}{(2)^3} \det \left[2 \begin{bmatrix} 2 & 3 & 2 & 1 \\ 1 & 2 & 1 & 2 \\ 1 & -1 & -1 & 3 \\ 2 & 1 & 1 & -2 \end{bmatrix} - \begin{bmatrix} 0 \\ 0 \\ 0 \\ 0 \end{bmatrix} [-1 \quad 1 \quad 1 \quad 2] \right]$$

or

$$\frac{1}{8} \det \begin{bmatrix} 4 & 6 & 4 & 2 \\ 2 & 4 & 2 & 4 \\ 2 & -2 & -2 & 6 \\ 4 & 2 & 2 & -4 \end{bmatrix}$$

The condensation process continues:

$$\frac{1}{8(4)^2} \det \left[4 \begin{bmatrix} 4 & 2 & 4 \\ -2 & -2 & 6 \\ 2 & 2 & -4 \end{bmatrix} - \begin{bmatrix} 2 \\ 2 \\ 4 \end{bmatrix} \begin{bmatrix} 6 & 4 & 2 \end{bmatrix} \right]$$

which is equal to

$$\frac{1}{128} \det \left[\begin{bmatrix} 16 & 8 & 16 \\ -8 & -8 & 24 \\ 8 & 8 & -16 \end{bmatrix} - \begin{bmatrix} 12 & 8 & 4 \\ 12 & 8 & 4 \\ 24 & 16 & 8 \end{bmatrix} \right]$$

$$= \frac{1}{128} \det \begin{bmatrix} 4 & 0 & 12 \\ -20 & -16 & 20 \\ -16 & -8 & -24 \end{bmatrix}$$

or

$$\frac{1}{128} (1536 + 0 + 1920 - 3072 - 0 + 640) = \frac{1}{128} (1024) = 8$$

2.9 EXERCISES

In all four of these exercises the objective is to evaluate the determinant by any method desired. If the determinant is equal to zero, the reason must be found and stated.

2.1 Find det A if

$$A = \begin{bmatrix} 1 & 0 & 3 \\ 2 & -1 & 4 \\ 3 & -2 & -1 \end{bmatrix}$$

2.2 Find det AB and det BA if A is as in exercise 2.1 and

$$B = \begin{bmatrix} 3 & 3 & 2 \\ 2 & 1 & -3 \\ 1 & -1 & -4 \end{bmatrix}$$

2.3 Find det AB and det BA if

$$A = \begin{bmatrix} 1 & 1 & 0 & 1 \\ 1 & 0 & 0 & 2 \\ -1 & 0 & 2 & -1 \\ 2 & 1 & 2 & 1 \\ 0 & 1 & 1 & 2 \end{bmatrix}$$

and

$$B = \begin{bmatrix} 1 & 0 & 1 & 0 & 1 \\ -1 & 1 & 2 & 1 & 0 \\ 1 & 2 & 1 & -1 & 2 \\ 2 & -1 & 0 & 1 & 0 \end{bmatrix}$$

2.4 Find det A if

$$A = \begin{bmatrix} 1 & 0 & 1 & 3 & -2 & 1 \\ 1 & -1 & 2 & -1 & 0 & 3 \\ 0 & 4 & -1 & 1 & 1 & 0 \\ 1 & 1 & -2 & -2 & 2 & 3 \\ 1 & 1 & 0 & 0 & -2 & 1 \\ 2 & 0 & 3 & 1 & 1 & 0 \end{bmatrix}$$

THREE
MATRIX INVERSION

3.1 INTRODUCTION

The nth-order set of simultaneous, linear algebraic equations in the n unknowns $x_1, x_2, x_3, \ldots, x_n$

$$a_{11}x_1 + a_{12}x_2 + a_{13}x_3 + \cdots a_{1n}x_n = y_1$$

$$a_{21}x_1 + a_{22}x_2 + a_{23}x_3 + \cdots a_{2n}x_n = y_2$$

$$a_{31}x_1 + a_{32}x_2 + a_{33}x_3 + \cdots a_{3n}x_n = y_3$$

$$\cdots$$

$$a_{n1}x_1 + a_{n2}x_2 + a_{n3}x_3 + \cdots a_{nn}x_n = y_n$$

can conveniently be represented by the matrix equation

$$\begin{bmatrix} a_{11} & a_{12} & a_{13} & \cdots & a_{1n} \\ a_{21} & a_{22} & a_{23} & \cdots & a_{2n} \\ a_{31} & a_{32} & a_{33} & \cdots & a_{3n} \\ & & \cdots & & \\ a_{n1} & a_{n2} & a_{n3} & \cdots & a_{nn} \end{bmatrix} \begin{bmatrix} x_1 \\ x_2 \\ x_3 \\ \cdots \\ x_n \end{bmatrix} = \begin{bmatrix} y_1 \\ y_2 \\ y_3 \\ \cdots \\ y_n \end{bmatrix}$$

or more simply by

$$AX = Y \tag{3.1}$$

where A, as indicated in Section 1.2, is a square matrix of coefficients having elements a_{ij} and where X and Y are $n \times 1$ column vectors with elements x_i and y_i, respectively.

Because division of matrices is not permitted, one method for the solution of matrix equations such as the one shown in Eq. (3.1) is called *matrix inversion*.

3.2 MATRIX INVERSION

If Eq. (3.1) is premultiplied by an $n \times n$ square matrix B so that

$$BAX = BY$$

a solution for the unknowns, X, will evolve if the product BA is equal to the identity matrix I:

$$BAX = IX = BY$$

or

$$X = BY \tag{3.2}$$

If

$$BA = AB = I$$

the matrix B is said to be the inverse of A:

$$B = A^{-1} \tag{3.3a}$$

and, of course, the inverse of the inverse is the matrix itself:

$$A = B^{-1} \tag{3.3b}$$

or

$$(A^{-1})^{-1} = A$$

It should be recalled that, in general, matrix multiplication is not commutative. The multiplication of a matrix by its inverse is one specific case in which matrix multiplication is commutative:

$$AA^{-1} = A^{-1}A = I \tag{3.4}$$

3.3 PROPERTIES OF THE INVERSE

The inverse of a product of two matrices is the product of the inverses taken in reverse order. This is easily proved. Consider the product AB and postmultiply by $B^{-1}A^{-1}$. Because matrix multiplication is associative, this product can be taken with a rearrangement of the parentheses and then by a straightforward application of the definition of the matrix inverse

$$AB(B^{-1}A^{-1}) = A(BB^{-1}A^{-1}) = AIA^{-1} = AA^{-1} = I$$

In addition, the inverse of the transpose of a matrix is equal to the transpose of its inverse

$$(A^{\mathrm{T}})^{-1} = (A^{-1})^{\mathrm{T}} \tag{3.5}$$

negative powers of a matrix are related to its inverse

$$A^{-n} = (A^{-1})^n \tag{3.6}$$

and the determinant of the product of a matrix and its inverse must be equal to unity

$$\det (AA^{-1}) = \det I = 1 \tag{3.7}$$

If the matrix does not possess an inverse, it is said to be singular, but if a matrix does possess an inverse, the inverse is unique. Consider that

$$AB = BA = I$$

which indicates $B = A^{-1}$, and pose that

$$AC = CA = I$$

Then

$$B = BI = B(AC) = (BA)C = IC = C$$

which shows that a matrix cannot have more than one inverse.

In addition, if $AB = 0$, then either $A = 0$ or $B = 0$ or both A and B are singular. Recall the discusssion in Section 1.7, which dealt with matrix multiplication. There, in an example, it is shown that two 3×3 matrices $A \neq 0$ and $B \neq 0$ yield a product $AB = 0$. In this case, however, both A and B are singular, which is demonstrated in one of the examples in Section 3.5.

Finally, if A, B, and C are all $n \times n$ and A is nonsingular, then $AB = AC$ implies that $B = C$. To show this, consider

$$AB = AC$$

and premultiply both sides by A^{-1}:

$$A^{-1}AB = IB = A^{-1}AC = IC$$

or

$$B = C$$

Thus there is a cancellation law in matrix multiplication over the set of all $n \times n$ nonsingular matrices.

3.4 THE ADJOINT MATRIX

The *adjoint matrix*, sometimes called the *adjugate matrix* and in this development referred to merely as the *adjoint*, applies only to a square matrix and is the transpose of the cofactor matrix:

$$\text{adj } A = (A^C)^T \tag{3.8}$$

Because symmetrical matrices possess symmetrical cofactor matrices, the adjoint of a symmetrical matrix is the cofactor matrix itself.

The matrix that is of nth order:

$$A = \begin{bmatrix} a_{11} & a_{12} & a_{13} & \cdots & a_{1n} \\ a_{21} & a_{22} & a_{23} & \cdots & a_{2n} \\ a_{31} & a_{32} & a_{33} & \cdots & a_{3n} \\ & & \cdots & & \\ a_{n1} & a_{n2} & a_{n3} & \cdots & a_{nn} \end{bmatrix}$$

has been observed to possess a cofactor matrix:

$$A^C = \begin{bmatrix} A_{11} & A_{12} & A_{13} & \cdots & A_{1n} \\ A_{21} & A_{22} & A_{23} & \cdots & A_{2n} \\ A_{31} & A_{32} & A_{33} & \cdots & A_{3n} \\ & & \cdots & & \\ A_{n1} & A_{n2} & A_{n3} & \cdots & A_{nn} \end{bmatrix}$$

and it has an adjoint:

$$\text{adj } A = \begin{bmatrix} A_{11} & A_{21} & A_{31} & \cdots & A_{n1} \\ A_{12} & A_{22} & A_{32} & \cdots & A_{n2} \\ A_{13} & A_{23} & A_{33} & \cdots & A_{n3} \\ & & \cdots & & \\ A_{1n} & A_{2n} & A_{3n} & \cdots & A_{nn} \end{bmatrix}$$

3.5 DETERMINATION OF THE INVERSE

Suppose an $n \times n$ matrix A is postmultiplied by its adjoint and that the product is designated as P:

$$A(\text{adj } A) = \begin{bmatrix} a_{11} & a_{12} & a_{13} & \cdots & a_{1n} \\ a_{21} & a_{22} & a_{23} & \cdots & a_{2n} \\ a_{31} & a_{32} & a_{33} & \cdots & a_{3n} \\ & & \cdots & & \\ a_{n1} & a_{n2} & a_{n3} & \cdots & a_{nn} \end{bmatrix} \begin{bmatrix} A_{11} & A_{21} & A_{31} & \cdots & A_{n1} \\ A_{12} & A_{22} & A_{32} & \cdots & A_{n2} \\ A_{13} & A_{23} & A_{33} & \cdots & A_{n3} \\ & & \cdots & & \\ A_{1n} & A_{2n} & A_{3n} & \cdots & A_{nn} \end{bmatrix} = P$$

The elements of P may be divided into two categories; those that lie upon its principal diagonal:

$$p_{11} = a_{11}A_{11} + a_{12}A_{12} + a_{13}A_{13} + \cdots + a_{1n}A_{1n}$$

$$p_{22} = a_{21}A_{21} + a_{22}A_{22} + a_{23}A_{23} + \cdots + a_{2n}A_{2n}$$

$$p_{33} = a_{31}A_{31} + a_{32}A_{32} + a_{33}A_{33} + \cdots + a_{3n}A_{3n}$$

$$\cdots$$

$$p_{nn} = a_{n1}A_{n1} + a_{n2}A_{n2} + a_{n3}A_{n3} + \cdots + a_{nn}A_{nn}$$

and those that do not:

$$p_{12} = a_{11}A_{21} + a_{12}A_{22} + a_{13}A_{23} + \cdots + a_{1n}A_{2n}$$

$$p_{13} = a_{11}A_{31} + a_{12}A_{32} + a_{13}A_{33} + \cdots + a_{1n}A_{3n}$$

$$p_{21} = a_{21}A_{11} + a_{22}A_{12} + a_{23}A_{13} + \cdots + a_{2n}A_{1n}$$

$$\cdots$$

$$p_{32} = a_{31}A_{21} + a_{32}A_{22} + a_{33}A_{23} + \cdots + a_{3n}A_{2n}$$

$$\cdots$$

$$p_{n3} = a_{n1}A_{31} + a_{n2}A_{32} + a_{n3}A_{33} + \cdots + a_{nn}A_{3n}$$

The elements of P that lie on the principal diagonal are all equal to the determinant of A (see Rule Twelve in Section 2.7 and recognize the Laplace expansion for the evaluation of a determinant). Observe that all other elements of P involve an expansion of one row of the matrix A with the cofactors of an entirely different row of A and are equal to zero (see

Rule Fourteen, Section 2.7). Thus the product of A and its adjoint is

$$A(\text{adj } A) = \begin{bmatrix} |A| & 0 & 0 & \cdots & 0 \\ 0 & |A| & 0 & \cdots & 0 \\ 0 & 0 & |A| & \cdots & 0 \\ & & \cdots & & \\ 0 & 0 & 0 & \cdots & |A| \end{bmatrix} = |A|I$$

If this is put into the form

$$A \frac{\text{adj } A}{\det A} = I$$

and then compared with Eq. (3.4)

$$AA^{-1} = I$$

it becomes evident that the inverse of the matrix A is equal to its adjoint divided by its determinant:

$$A^{-1} = \frac{\text{adj } A}{\det A} \tag{3.9}$$

Observe that if $\det A = 0$, the inverse of A cannot exist and A is therefore singular. Thus the condition necessary to make a matrix A singular is for $\det A$ to equal zero.

Example: The symmetrical matrix

$$A = \begin{bmatrix} 3 & -2 & 0 \\ -2 & 4 & -1 \\ 0 & -1 & 6 \end{bmatrix}$$

has been shown in an example in Section 2.6 to possess a determinant

$$\det A = 45$$

and a cofactor matrix

$$A^c = \begin{bmatrix} 23 & 12 & 2 \\ 12 & 18 & 3 \\ 2 & 3 & 8 \end{bmatrix}$$

that is symmetrical. This makes the adjoint that is equal to the transpose of the cofactor matrix equal to the cofactor matrix, and then, in accordance with Eq. (3.9), the inverse of A is

$$A^{-1} = \begin{bmatrix} 23/45 & 4/15 & 2/45 \\ 4/15 & 2/5 & 1/15 \\ 2/45 & 1/15 & 8/45 \end{bmatrix}$$

which is also observed to be symmetrical.

Thus symmetrical matrices possess

(a) symmetrical transposes
(b) symmetrical cofactor matrices
(c) symmetrical adjoints
(d) symmetrical inverses

The evaluation of the inverse can always be concluded with a check on its validity via Eq. (3.4). In the example just concluded

$$\begin{bmatrix} 23/45 & 4/15 & 2/45 \\ 4/15 & 2/5 & 1/15 \\ 2/45 & 1/15 & 8/45 \end{bmatrix} \begin{bmatrix} 3 & -2 & 0 \\ -2 & 4 & -1 \\ 0 & -1 & 6 \end{bmatrix} = \begin{bmatrix} 1 & 0 & 0 \\ 0 & 1 & 0 \\ 0 & 0 & 1 \end{bmatrix}$$

Example: In this example the matrix A is not symmetrical:

$$A = \begin{bmatrix} 1 & 2 & 3 \\ 0 & 4 & -1 \\ 1 & 1 & -2 \end{bmatrix}$$

It has a determinant

$$\det A = -8 - 2 + 0 - 12 + 1 - 0 = -22 + 1 = -21$$

and nine cofactors

$$A_{11} = + \begin{vmatrix} 4 & -1 \\ 1 & -2 \end{vmatrix} = (-8 + 1) = -7$$

$$A_{12} = - \begin{vmatrix} 0 & -1 \\ 1 & -2 \end{vmatrix} = -1$$

$$A_{13} = + \begin{vmatrix} 0 & 4 \\ 1 & 1 \end{vmatrix} = -4$$

$$A_{21} = - \begin{vmatrix} 2 & 3 \\ 1 & -2 \end{vmatrix} = -(-4 - 3) = -(-7) = 7$$

$$A_{22} = + \begin{vmatrix} 1 & 3 \\ 1 & -2 \end{vmatrix} = -2 - 3 = -5$$

$$A_{23} = - \begin{vmatrix} 1 & 2 \\ 1 & 1 \end{vmatrix} = -(1 - 2) = -(-1) = 1$$

$$A_{31} = + \begin{vmatrix} 2 & 3 \\ 4 & -1 \end{vmatrix} = -2 - 12 = -14$$

$$A_{32} = - \begin{vmatrix} 1 & 3 \\ 0 & -1 \end{vmatrix} = -(-1) = 1$$

and

$$A_{33} = + \begin{vmatrix} 1 & 2 \\ 0 & 4 \end{vmatrix} = 4$$

The cofactor matrix is

$$A^c = \begin{bmatrix} -7 & -1 & -4 \\ 7 & -5 & 1 \\ -14 & 1 & 4 \end{bmatrix}$$

and the adjoint is the transpose of the cofactor matrix is

$$\text{adj } A = \begin{bmatrix} -7 & 7 & -14 \\ -1 & -5 & 1 \\ -4 & 1 & 4 \end{bmatrix}$$

All the foregoing makes the inverse

$$A^{-1} = \frac{\text{adj } A}{\det A} = \begin{bmatrix} 1/3 & -1/3 & 2/3 \\ 1/21 & 5/21 & -1/21 \\ 4/21 & -1/21 & -4/21 \end{bmatrix}$$

and this can be checked:

$$A^{-1}A = \begin{bmatrix} 1/3 & -1/3 & 2/3 \\ 1/21 & 5/21 & -1/21 \\ 4/21 & -1/21 & -4/21 \end{bmatrix} \begin{bmatrix} 1 & 2 & 3 \\ 0 & 4 & -1 \\ 1 & 1 & -2 \end{bmatrix} = \begin{bmatrix} 1 & 0 & 0 \\ 0 & 1 & 0 \\ 0 & 0 & 1 \end{bmatrix}$$

Example: The nonsymmetrical second-order matrix

$$A = \begin{bmatrix} 4 & -1 \\ 1 & 6 \end{bmatrix}$$

has a determinant

$$\det A = 24 + 1 = 25$$

and a cofactor matrix

$$A^c = \begin{bmatrix} 6 & -1 \\ 1 & 4 \end{bmatrix}$$

and an adjoint

$$\text{adj } A = \begin{bmatrix} 6 & 1 \\ -1 & 4 \end{bmatrix}$$

Its inverse is

$$A^{-1} = \frac{\text{adj } A}{\det A} = \begin{bmatrix} 6/25 & 1/25 \\ -1/25 & 4/25 \end{bmatrix}$$

This is easily checked:

$$\begin{bmatrix} 4 & -1 \\ 1 & 6 \end{bmatrix} \begin{bmatrix} 6/25 & 1/25 \\ -1/25 & 4/25 \end{bmatrix} = \begin{bmatrix} 1 & 0 \\ 0 & 1 \end{bmatrix}$$

and it is observed that the inverse of a second-order matrix is obtained by swapping the elements that lie on the principal diagonal, multiplying the off-diagonal elements by -1, and then dividing all elements by the determinant.

Example: In Section 1.7 it is shown that two matrices

$$A = \begin{bmatrix} 0 & 1 & 2 \\ 0 & -1 & 2 \\ 0 & 3 & 2 \end{bmatrix}$$

and

$$B = \begin{bmatrix} 3 & 2 & -1 \\ 0 & 0 & 0 \\ 0 & 0 & 0 \end{bmatrix}$$

can be multiplied to yield a null matrix

$$AB = \begin{bmatrix} 0 & 1 & 2 \\ 0 & -1 & 2 \\ 0 & 3 & 5 \end{bmatrix} \begin{bmatrix} 3 & 2 & -1 \\ 0 & 0 & 0 \\ 0 & 0 & 0 \end{bmatrix} = \begin{bmatrix} 0 & 0 & 0 \\ 0 & 0 & 0 \\ 0 & 0 & 0 \end{bmatrix}$$

even though neither A nor B is null. Both A and B are singular, however, because $\det A = \det B = 0$.

3.6 A PROOF OF CRAMER'S RULE

The procedure employed to derive an expression for the inverse of a matrix can be used to prove Cramer's rule. Look at the set of n simultaneous, linear algebraic equations in n unknowns:

$$a_{11}x_1 + a_{12}x_2 + a_{13}x_3 + \cdots + a_{1n}x_n = b_1$$
$$a_{21}x_1 + a_{22}x_2 + a_{23}x_3 + \cdots + a_{2n}x_n = b_2$$
$$a_{31}x_1 + a_{32}x_3 + a_{33}x_3 + \cdots + a_{3n}x_n = b_3$$
$$\cdots$$
$$a_{n1}x_1 + a_{n2}x_2 + a_{n3}x_3 + \cdots + a_{nn}x_n = b_n$$

and put them in the matrix form $AX = B$:

$$
\begin{bmatrix}
a_{11} & a_{12} & a_{13} & \cdots & a_{1n} \\
a_{21} & a_{22} & a_{23} & \cdots & a_{2n} \\
a_{31} & a_{32} & a_{33} & \cdots & a_{3n} \\
& & \cdots & & \\
a_{n1} & a_{n2} & a_{n3} & \cdots & a_{nn}
\end{bmatrix}
\begin{bmatrix}
x_1 \\
x_2 \\
x_3 \\
\cdots \\
x_n
\end{bmatrix}
=
\begin{bmatrix}
b_1 \\
b_2 \\
b_3 \\
\cdots \\
b_n
\end{bmatrix}
$$

To solve for the vector X, one takes the inverse so that $X = A^{-1}B$. Using the definition of the inverse, Eq. (3.9), X can be represented by

$$X = \frac{\text{adj } A}{\det A} B$$

or

$$(\det A)X = (\text{adj } A)B$$

This representation may be displayed showing the elements of A, X, and B:

$$
\det A
\begin{bmatrix}
x_1 \\
x_2 \\
x_3 \\
\cdots \\
x_n
\end{bmatrix}
=
\begin{bmatrix}
A_{11} & A_{21} & A_{31} & \cdots & A_{n1} \\
A_{12} & A_{22} & A_{32} & \cdots & A_{n2} \\
A_{13} & A_{23} & A_{33} & \cdots & A_{n3} \\
& & \cdots & & \\
A_{1n} & A_{2n} & A_{3n} & \cdots & A_{nn}
\end{bmatrix}
\begin{bmatrix}
b_1 \\
b_2 \\
b_3 \\
\cdots \\
b_n
\end{bmatrix}
$$

and when it is expanded one finds that

$$(\det A)x_1 = A_{11}b_1 + A_{21}b_2 + A_{31}b_3 + \cdots + A_{n1}b_n$$

$$(\det A)x_2 = A_{12}b_1 + A_{22}b_2 + A_{32}b_3 + \cdots + A_{n2}b_n$$

$$(\det A)x_3 = A_{13}b_1 + A_{23}b_2 + A_{33}b_3 + \cdots + A_{n3}b_n$$

$$\cdots$$

$$(\det A)x_n = A_{1n}b_1 + A_{2n}b_2 + A_{3n}b_3 + \cdots + A_{nn}b_n$$

In each of the foregoing equations the expression on the right side of the equal sign is an expansion of the elements b and the cofactors belonging to each column of A. For example, the first equation provides

$$(\det A)x_1 = b_1 A_{11} + b_2 A_{21} + b_3 A_{31} + \cdots + b_n A_{n1}$$

and the solution for x_1 may be written as

$$x_1 = \frac{\det (A_1)}{\det A}$$

where

$$
\det A_1 =
\begin{vmatrix}
b_1 & a_{12} & a_{13} & \cdots & a_{1n} \\
b_2 & a_{22} & a_{23} & \cdots & a_{2n} \\
b_3 & a_{32} & a_{33} & \cdots & a_{3n} \\
& & \cdots & & \\
b_n & a_{n2} & a_{n3} & \cdots & a_{nn}
\end{vmatrix}
$$

It is easy to see that for x_2

$$x_2 = \frac{\det (A_2)}{\det A}$$

where

$$
\det A_2 =
\begin{vmatrix}
a_{11} & b_1 & a_{13} & \cdots & a_{1n} \\
a_{21} & b_2 & a_{23} & \cdots & a_{2n} \\
a_{31} & b_3 & a_{33} & \cdots & a_{3n} \\
& & \cdots & & \\
a_{n1} & b_n & a_{n3} & \cdots & a_{nn}
\end{vmatrix}
$$

Continuation of this reasoning provides the "official" statement of Cramer's rule: the system of n simultaneous, linear algebraic equations in

n unknowns, $AX = B$ will have exactly one solution:

$$x_i = \frac{\det (A_i)}{\det A} \qquad i = 1, 2, 3, \cdots, n \qquad (3.10)$$

provided that $\det A \neq 0$ and where $\det (A_i)$ is the determinant of A with the ith column of elements replaced by the n elements of B.

3.7 METHODS FOR OBTAINING THE INVERSE

Six methods for obtaining the inverse of a matrix are considered in this book:

1. The method used in Section 3.5, which comes from the fundamental definition of the inverse given by Eq. (3.9).
2. The method that employs a sequence of elementary transformations, which is discussed and illustrated in Section 3.9.
3. The so-called sweep-out method, which is discussed and illustrated in Section 3.10.
4. The method that utilizes partitioned matrices, which is presented in Chapter 4.
5. The method of Leverrier, which is presented in Chapter 7.
6. The method that involves an application of Cayley-Hamilton theorem, which is presented in Chapter 8.

3.8 ELEMENTARY TRANSFORMATION MATRICES

Three elementary transformation matrices can be used to perform a variety of tasks in the realm of matrix adjustment and simplification. One of the most useful applications of these elementary transformation matrices is in an efficient computational scheme that can be employed to find the inverse of any matrix in general, and those of fourth and higher order in particular.

Before beginning a detailed discussion of the three elementary transformation matrices, all of which are designated by the letter E with a subscript, it is important to stress that the elementary transformation matrices make adjustments to the rows and columns of $n \times n$ square matrices and to the rows of conformable rectangular matrices. In the case of square matrices, premultiplication of a matrix by an elementary transformation matrix makes an adjustment to a row. Postmultiplication of a matrix by an elementary transformation matrix makes an adjustment to a column.

The elementary transformation matrix of the first kind which is designated as E_1 multiplies a row or column of a matrix by a constant. This matrix is also called the *scaling transformation matrix* and adjusts the identity matrix by placing the numeral 1 at the appropriate row or column by the multiplier.

Example: For the matrix A

$$A = \begin{bmatrix} 1 & 2 & 3 & 4 \\ -1 & 5 & -6 & 7 \\ 3 & 4 & 5 & -6 \\ 1 & 0 & 2 & 0 \end{bmatrix}$$

if it is desired to multiply the second row by 2, select the elementary transformation matrix E_1 as

$$E_1 = \begin{bmatrix} 1 & 0 & 0 & 0 \\ 0 & 2 & 0 & 0 \\ 0 & 0 & 1 & 0 \\ 0 & 0 & 0 & 1 \end{bmatrix}$$

and premultiply A to obtain

$$E_1 A = \begin{bmatrix} 1 & 0 & 0 & 0 \\ 0 & 2 & 0 & 0 \\ 0 & 0 & 1 & 0 \\ 0 & 0 & 0 & 0 \end{bmatrix} \begin{bmatrix} 1 & 2 & 3 & 4 \\ -1 & 5 & -6 & 7 \\ 3 & 4 & 5 & -6 \\ 1 & 0 & 2 & 0 \end{bmatrix} = \begin{bmatrix} 1 & 2 & 3 & 4 \\ -2 & 10 & -12 & 14 \\ 3 & 4 & 5 & -6 \\ 1 & 0 & 2 & 0 \end{bmatrix}$$

If it is desired to multiply the third column by 4, use

$$E_1 = \begin{bmatrix} 1 & 0 & 0 & 0 \\ 0 & 1 & 0 & 0 \\ 0 & 0 & 4 & 0 \\ 0 & 0 & 0 & 1 \end{bmatrix}$$

and postmultiply A to obtain

$$AE_1 = \begin{bmatrix} 1 & 2 & 3 & 4 \\ -1 & 5 & -6 & 7 \\ 3 & 4 & 5 & -6 \\ 1 & 0 & 2 & 0 \end{bmatrix} \begin{bmatrix} 1 & 0 & 0 & 0 \\ 0 & 1 & 0 & 0 \\ 0 & 0 & 4 & 0 \\ 0 & 0 & 0 & 1 \end{bmatrix} = \begin{bmatrix} 1 & 2 & 12 & 4 \\ -1 & 5 & -24 & 7 \\ 3 & 4 & 20 & -6 \\ 1 & 0 & 8 & 0 \end{bmatrix}$$

The elementary transformation matrix of the second kind is sometimes called the *interchanging transformation matrix* and is designated by E_2. It accomplishes a row or column switch and is also derived from the identity matrix. The switch or interchange is accomplished by swapping the elements of unity in the identity matrix at the point where the interchange is to be effected.

Example: For the matrix A

$$A = \begin{bmatrix} 1 & -2 & 3 \\ 0 & -1 & 2 \\ 3 & 1 & 5 \end{bmatrix}$$

an interchange of rows one and two may be accomplished by premultiplying A

$$E_2 A = \begin{bmatrix} 0 & 1 & 0 \\ 1 & 0 & 0 \\ 0 & 0 & 1 \end{bmatrix} \begin{bmatrix} 1 & -2 & 3 \\ 0 & -1 & 2 \\ 3 & 4 & 5 \end{bmatrix} = \begin{bmatrix} 0 & -1 & 2 \\ 1 & -2 & 3 \\ 3 & 4 & 5 \end{bmatrix}$$

An interchange of columns one and three involves a postmultiplication

$$AE_2 = \begin{bmatrix} 1 & -2 & 3 \\ 0 & -1 & 2 \\ 3 & 4 & 5 \end{bmatrix} \begin{bmatrix} 0 & 0 & 1 \\ 0 & 1 & 0 \\ 1 & 0 & 0 \end{bmatrix} = \begin{bmatrix} 3 & -2 & 1 \\ 2 & -1 & 0 \\ 5 & 4 & 3 \end{bmatrix}$$

The elementary transformation of the third kind is often referred to as the *combining transformation matrix* and is designated by E_3. Its employment permits the addition of a scalar or constant multiple of one row or column to another row or column where the row or column that is multiplied remains unchanged. As in the cases of the first and second elementary transformation matrices, rows are handled by premultiplications, while postmultiplications involve columns.

Here, too, the identity matrix is used as the basis for the elementary transformation matrices of the third kind. However, the placement of the elements in the identity matrix requires careful consideration.

Consider first a row operation in which a row is to be multiplied by a scalar and then added to another row and where, in the eventual result, the row that is multiplied is left unchanged. In this case the row in the identity matrix that corresponds to the row that is to be left unchanged is left strictly alone. If the ith row is to be multiplied by the scalar with the intent of operating on the jth row, the scalar is placed in the jith position in the identity matrix.

Example: Suppose in

$$A = \begin{bmatrix} 1 & -2 & 3 \\ 0 & -1 & 2 \\ 3 & 4 & 5 \end{bmatrix}$$

it is desired to subtract four times the second row from the third row, leaving the second row unchanged. In this case the identity matrix will have no changes in the first and second rows, and the scalar, -4, is placed in the second column of the third row because the second row is to remain unchanged and the third row is to be adjusted. This makes the premultiplier (premultiplier because this is a row operation)

$$E_3 = \begin{bmatrix} 1 & 0 & 0 \\ 0 & 0 & 1 \\ 0 & -4 & 1 \end{bmatrix}$$

and the actual operation is

$$E_3A = \begin{bmatrix} 1 & 0 & 0 \\ 0 & 1 & 0 \\ 0 & -4 & 1 \end{bmatrix}\begin{bmatrix} 1 & -2 & 3 \\ 0 & -1 & 2 \\ 3 & 4 & 5 \end{bmatrix} = \begin{bmatrix} 1 & -2 & 3 \\ 0 & -1 & 2 \\ 3 & 8 & -3 \end{bmatrix}$$

The procedure for columns through the use of a postmultiplier is similar. For the columns that are to remain unchanged, the corresponding columns of the identity matrix are left unchanged. If the ith column is to multiplied by the scalar and then added to the jth column leaving the jth column unchanged, the scalar is placed in the ijth position in the identity matrix.

Example: If in

$$A = \begin{bmatrix} 1 & -2 & 3 \\ 0 & -1 & 2 \\ 3 & 4 & 5 \end{bmatrix}$$

it is desired to multiply the first column by 2 and add the result to the third column, leaving the first column unchanged, the elementary transmission matrix is

$$E_3 = \begin{bmatrix} 1 & 0 & 2 \\ 0 & 1 & 0 \\ 0 & 0 & 1 \end{bmatrix}$$

and the actual operation is

$$AE_3 = \begin{bmatrix} 1 & -2 & 3 \\ 0 & -1 & 2 \\ 3 & 4 & 5 \end{bmatrix} \begin{bmatrix} 1 & 0 & 2 \\ 0 & 1 & 0 \\ 0 & 0 & 1 \end{bmatrix} = \begin{bmatrix} 1 & -2 & 5 \\ 0 & -1 & 2 \\ 3 & 4 & 11 \end{bmatrix}$$

Successive transformations of a given matrix may be executed by a single transformation matrix T, which is the product of all corresponding elementary transformation matrices in the proper order. This is a useful and computationally efficient procedure that can be employed in the determination of matrix inverses.

Suppose it is necessary to employ n elementary transformation matrices to effect the desired transformation of matrix A to matrix B. One selects an individual transformation matrix $T_i = E_{1i}$, E_{2i}, or E_{3i} for $i = 1, 2, 3,$ $\cdots$, n. The transformation matrix T will then be for a row operation sequence

$$T = T_n T_{n-1} \cdots T_3 T_2 T_1$$

and

$$B = TA$$

If it is desired to use a column operation sequence, then

$$T = T_1 T_2 T_3 \cdots T_{n-1} T_n$$

and

$$B = AT$$

Example: Suppose that it is desired to transform

$$A = \begin{bmatrix} 2 & 3 & -1 \\ 4 & -1 & 5 \\ 0 & 2 & 6 \end{bmatrix}$$

into

$$B = \begin{bmatrix} 1 & b_{12} & b_{13} \\ 0 & b_{22} & b_{23} \\ 0 & b_{32} & b_{33} \end{bmatrix}$$

using no more than three elementary transformation matrices.

Step One: Use the scaling transformation matrix (the elementary transformation matrix of the first kind) to put a 1 into position b_{11}. Here, $T_1 =$

E_{11} and $B_1 = T_1A$

$$B_1 = \begin{bmatrix} 1/2 & 0 & 0 \\ 0 & 1 & 0 \\ 0 & 0 & 1 \end{bmatrix} \begin{bmatrix} 2 & 3 & -1 \\ 4 & -1 & 5 \\ 0 & 2 & 6 \end{bmatrix} = \begin{bmatrix} 1 & 3/2 & -1/2 \\ 4 & -1 & 5 \\ 0 & 2 & 6 \end{bmatrix}$$

Step Two: Multiply row one of B_1 by -4 and add the result to row two, leaving row one unchanged. This involves the use of the combining transformation matrix (the elementary transformation matrix of the third kind). Here, $T_2 = E_{32}$ and $B_2 = T_2B_1$ and the proper placement of the -4 in T_2 should be carefully noted:

$$B_2 = \begin{bmatrix} 1 & 0 & 0 \\ -4 & 1 & 0 \\ 0 & 0 & 1 \end{bmatrix} \begin{bmatrix} 1 & 3/2 & -1/2 \\ 4 & -1 & 5 \\ 0 & 2 & 6 \end{bmatrix} = \begin{bmatrix} 1 & 3/2 & -1/2 \\ 0 & -7 & 7 \\ 0 & 2 & 6 \end{bmatrix}$$

Further steps are not required, because the desired result has been achieved.

Before leaving this example, look at the product $T = T_2T_1$:

$$T = \begin{bmatrix} 1 & 0 & 0 \\ -4 & 1 & 0 \\ 0 & 0 & 1 \end{bmatrix} \begin{bmatrix} 1/2 & 0 & 0 \\ 0 & 1 & 0 \\ 0 & 0 & 1 \end{bmatrix} = \begin{bmatrix} 1/2 & 0 & 0 \\ -2 & 1 & 0 \\ 0 & 0 & 1 \end{bmatrix}$$

and observe that $B_2 = TA$

$$B_2 = \begin{bmatrix} 1/2 & 0 & 0 \\ -2 & 1 & 0 \\ 0 & 0 & 1 \end{bmatrix} \begin{bmatrix} 2 & 3 & -1 \\ 4 & -1 & 5 \\ 0 & 2 & 6 \end{bmatrix} = \begin{bmatrix} 1 & 3/2 & -1/2 \\ 0 & -7 & 7 \\ 0 & 2 & 6 \end{bmatrix}$$

which is identical to the previous result.

Example: If it is desired to employ column operations on the matrix A

$$A = \begin{bmatrix} 2 & 3 & -1 \\ 4 & -1 & 5 \\ 0 & 2 & 6 \end{bmatrix}$$

to form

$$B = \begin{bmatrix} 1 & 0 & 0 \\ b_{21} & b_{22} & b_{23} \\ b_{31} & b_{32} & b_{33} \end{bmatrix}$$

three steps are required, because row one contains no zeroes.

Step One: Postmultiply A with the transformation matrix $T_1 = E_{11}$ to obtain $B_1 = AT_1$, which puts a 1 in the b_{11} position.

$$B_1 = \begin{bmatrix} 2 & 3 & -1 \\ 4 & -1 & 5 \\ 0 & 2 & 6 \end{bmatrix} \begin{bmatrix} 1/2 & 0 & 0 \\ 0 & 1 & 0 \\ 0 & 0 & 1 \end{bmatrix} = \begin{bmatrix} 1 & 3 & -1 \\ 2 & -1 & 5 \\ 0 & 2 & 6 \end{bmatrix}$$

Step Two: To multiply column one by -3 and add the result to column two and, at the same time, leave column one unchanged, an elementary transformation matrix of the third kind is required. Thus $T_2 = E_{32}$ and $B_2 = B_1T_2$:

$$B_2 = \begin{bmatrix} 1 & 3 & -1 \\ 2 & -1 & 5 \\ 0 & 2 & 6 \end{bmatrix} \begin{bmatrix} 1 & -3 & 0 \\ 0 & 1 & 0 \\ 0 & 0 & 1 \end{bmatrix} = \begin{bmatrix} 1 & 0 & -1 \\ 2 & -7 & 5 \\ 0 & 2 & 6 \end{bmatrix}$$

Step Three: Use the elementary transformation matrix of the third kind to multiply column three by 1 and then add the result to column one. This yields $B_3 = B_2T_3$ where $T_3 = E_{33}$:

$$B_3 = \begin{bmatrix} 1 & 0 & -1 \\ 2 & -7 & 5 \\ 0 & 2 & 6 \end{bmatrix} \begin{bmatrix} 1 & 0 & 1 \\ 0 & 1 & 0 \\ 0 & 0 & 1 \end{bmatrix} = \begin{bmatrix} 1 & 0 & 0 \\ 2 & -7 & 7 \\ 0 & 2 & 6 \end{bmatrix}$$

This result may be confirmed by forming $T = T_1T_2T_3$ and then postmultiplying A by T. First get $T = T_1T_2T_3$

$$T = \begin{bmatrix} 1/2 & 0 & 0 \\ 0 & 1 & 0 \\ 0 & 0 & 1 \end{bmatrix} \begin{bmatrix} 1 & -3 & 0 \\ 0 & 1 & 0 \\ 0 & 0 & 1 \end{bmatrix} \begin{bmatrix} 1 & 0 & 1 \\ 0 & 1 & 0 \\ 0 & 0 & 1 \end{bmatrix} = \begin{bmatrix} 1/2 & 0 & 0 \\ 0 & 1 & 0 \\ 0 & 0 & 1 \end{bmatrix} \begin{bmatrix} 1 & -3 & 1 \\ 0 & 1 & 0 \\ 0 & 0 & 1 \end{bmatrix}$$

or

$$T = \begin{bmatrix} 1/2 & -3/2 & 1/2 \\ 0 & 1 & 0 \\ 0 & 0 & 1 \end{bmatrix}$$

The result is confirmed when $B = AT$ is formed:

$$B = \begin{bmatrix} 2 & 3 & -1 \\ 4 & -1 & 5 \\ 0 & 2 & 6 \end{bmatrix} \begin{bmatrix} 1/2 & -3/2 & 1/2 \\ 0 & 1 & 0 \\ 0 & 0 & 1 \end{bmatrix} = \begin{bmatrix} 1 & 0 & 0 \\ 2 & -7 & 7 \\ 0 & 2 & 6 \end{bmatrix}$$

3.9 INVERSE OF A MATRIX BY SUCCESSIVE TRANSFORMATIONS

The inverse of a matrix can be obtained through the use of successive applications of the elementary transformation matrices. The procedure is identical to that employed in the examples in Section 3.8. Either row or column transformations may be used, but the two may not be mixed in a particular matrix inverse computation.

Example: In Section 3.5 the inverse of the matrix

$$A = \begin{bmatrix} 1 & 2 & 3 \\ 0 & 4 & -1 \\ 1 & 1 & -2 \end{bmatrix}$$

was obtained. It can be determined from successive applications of the elementary transformation matrices.

Step One: Get A_1 from $T_1 A$ where $T_1 = E_{31}$ involves a subtraction of row three from row one, with row one remaining unchanged:

$$T_1 = \begin{bmatrix} 1 & 0 & 0 \\ 0 & 1 & 0 \\ -1 & 0 & 1 \end{bmatrix}$$

and

$$A_1 = T_1 A = \begin{bmatrix} 1 & 0 & 0 \\ 0 & 1 & 0 \\ -1 & 0 & 1 \end{bmatrix} \begin{bmatrix} 1 & 2 & 3 \\ 0 & 4 & -1 \\ 1 & 1 & -2 \end{bmatrix} = \begin{bmatrix} 1 & 2 & 3 \\ 0 & 4 & -1 \\ 0 & -1 & -5 \end{bmatrix}$$

Step Two: Obtain A_2 from A_1 using $T_2 = E_{32}$. The use of T_2 multiplies row three by 2 and adds the result to row one, leaving row three unchanged:

$$T_2 = \begin{bmatrix} 1 & 0 & 2 \\ 0 & 1 & 0 \\ 0 & 0 & 1 \end{bmatrix}$$

and

$$A_2 = T_2 A_1 = \begin{bmatrix} 1 & 0 & 2 \\ 0 & 1 & 0 \\ 0 & 0 & 1 \end{bmatrix} \begin{bmatrix} 1 & 2 & 3 \\ 0 & 4 & -1 \\ 0 & -1 & -5 \end{bmatrix} = \begin{bmatrix} 1 & 0 & -7 \\ 0 & 4 & -1 \\ 0 & -1 & -5 \end{bmatrix}$$

At this point, to assist in the bookkeeping, it is useful to find $T_a = T_2 T_1$:

$$T_a = T_2 T_1 = \begin{bmatrix} 1 & 0 & 2 \\ 0 & 1 & 0 \\ 0 & 0 & 1 \end{bmatrix} \begin{bmatrix} 1 & 0 & 0 \\ 0 & 1 & 0 \\ -1 & 0 & 1 \end{bmatrix} = \begin{bmatrix} -1 & 0 & 2 \\ 0 & 1 & 0 \\ -1 & 0 & 1 \end{bmatrix}$$

Step Three: The scaling transformation $T_3 = E_{13}$ can be used to change the 4 to a 1 in the a_{22} position of A_2:

$$T_3 = \begin{bmatrix} 1 & 0 & 0 \\ 0 & 1/4 & 0 \\ 0 & 0 & 1 \end{bmatrix}$$

and

$$A_3 = T_3 A_2 = \begin{bmatrix} 1 & 0 & 0 \\ 0 & 1/4 & 0 \\ 0 & 1 & 0 \end{bmatrix} \begin{bmatrix} 1 & 0 & -7 \\ 0 & 4 & -1 \\ 0 & -1 & -5 \end{bmatrix} = \begin{bmatrix} 1 & 0 & -7 \\ 0 & 1 & -1/4 \\ 0 & -1 & -5 \end{bmatrix}$$

Step Four: Add row three to row two, leaving row two unchanged. This is accomplished by making $T_4 = E_{34}$, and the operation involves finding $A_4 = T_4 A_3$:

$$T_4 = \begin{bmatrix} 1 & 0 & 0 \\ 0 & 1 & 0 \\ 0 & 1 & 1 \end{bmatrix}$$

and

$$A_4 = T_4 A_3 = \begin{bmatrix} 1 & 0 & 0 \\ 0 & 1 & 0 \\ 0 & 1 & 1 \end{bmatrix} \begin{bmatrix} 1 & 0 & -7 \\ 0 & 1 & -1/4 \\ 0 & -1 & -5 \end{bmatrix} = \begin{bmatrix} 1 & 0 & -7 \\ 0 & 1 & -1/4 \\ 0 & 0 & -21/4 \end{bmatrix}$$

Now the bookkeeping requires that $T_b = T_4 T_3$:

$$T_b = T_4 T_3 = \begin{bmatrix} 1 & 0 & 0 \\ 0 & 1 & 0 \\ 0 & 1 & 1 \end{bmatrix} \begin{bmatrix} 1 & 0 & 0 \\ 0 & 1/4 & 0 \\ 0 & 0 & 1 \end{bmatrix} = \begin{bmatrix} 1 & 0 & 0 \\ 0 & 1/4 & 0 \\ 0 & 1/4 & 1 \end{bmatrix}$$

and $T_c = T_b T_a$

$$T_c = T_b T_a = \begin{bmatrix} 1 & 0 & 0 \\ 0 & 1/4 & 0 \\ 0 & 1/4 & 1 \end{bmatrix} \begin{bmatrix} -1 & 0 & 2 \\ 0 & 1 & 0 \\ -1 & 0 & 1 \end{bmatrix} = \begin{bmatrix} -1 & 0 & 2 \\ 0 & 1/4 & 0 \\ -1 & 1/4 & 1 \end{bmatrix}$$

Step Five: The time has come for a shot at column three. The first entry at the top of this column can be set equal to zero if row three is multiplied by $-4/3$ and then added to row one, leaving row three unchanged. This is accomplished by letting $T_5 = E_{35}$ so that $A_5 = T_5A_4$:

$$T_5 = \begin{bmatrix} 1 & 0 & -4/3 \\ 0 & 1 & 0 \\ 0 & 0 & 1 \end{bmatrix}$$

and

$$A_5 = T_5A_4 = \begin{bmatrix} 1 & 0 & -4/3 \\ 0 & 1 & 0 \\ 0 & 0 & 1 \end{bmatrix}\begin{bmatrix} 1 & 0 & -7 \\ 0 & 1 & -1/4 \\ 0 & 0 & -21/4 \end{bmatrix} = \begin{bmatrix} 1 & 0 & 0 \\ 0 & 1 & -1/4 \\ 0 & 0 & -21/4 \end{bmatrix}$$

Step Six: Multiplication of row three of A_5 by $-1/21$ with a subsequent addition of the result to row two, leaving row three unchanged in the next step. Thus $T_6 = E_{36}$

$$T_6 = \begin{bmatrix} 1 & 0 & 0 \\ 0 & 1 & -1/21 \\ 0 & 0 & 1 \end{bmatrix}$$

and $A_6 = T_6A_5$:

$$A_6 = T_6A_5 = \begin{bmatrix} 1 & 0 & 0 \\ 0 & 1 & -1/21 \\ 0 & 0 & 1 \end{bmatrix}\begin{bmatrix} 1 & 0 & 0 \\ 0 & 1 & -1/4 \\ 0 & 0 & -21/4 \end{bmatrix} = \begin{bmatrix} 1 & 0 & 0 \\ 0 & 1 & 0 \\ 0 & 0 & -21/4 \end{bmatrix}$$

Now for the bookkeeping. Let $T_d = T_6T_5$:

$$T_d = T_6T_5 = \begin{bmatrix} 1 & 0 & 0 \\ 0 & 1 & -1/21 \\ 0 & 0 & 1 \end{bmatrix}\begin{bmatrix} 1 & 0 & -4/3 \\ 0 & 1 & 0 \\ 0 & 0 & 1 \end{bmatrix} = \begin{bmatrix} 1 & 0 & -4/3 \\ 0 & 1 & -1/21 \\ 0 & 0 & 1 \end{bmatrix}$$

and with $T_e = T_dT_c$:

$$T_e = T_dT_c$$

$$= \begin{bmatrix} 1 & 0 & -4/3 \\ 0 & 1 & -1/21 \\ 0 & 0 & 1 \end{bmatrix}\begin{bmatrix} -1 & 0 & 2 \\ 0 & 1/4 & 0 \\ -1 & 1/4 & 1 \end{bmatrix} = \begin{bmatrix} 1/3 & -1/3 & 2/3 \\ 1/21 & 5/21 & -1/21 \\ -1 & 1/4 & 1 \end{bmatrix}$$

Step Seven: The final step consists of multiplying A_6 by the scaling transformation $T_7 = E_{17}$ to change the value of the single $-21/4$ term to unity:

$$T_7 = \begin{bmatrix} 1 & 0 & 0 \\ 0 & 1 & 0 \\ 0 & 0 & -4/21 \end{bmatrix}$$

With T_7 acting as a premultiplier for A_6, A_7 is the identity matrix:

$$A_7 = T_7 A_6 = \begin{bmatrix} 1 & 0 & 0 \\ 0 & 1 & 0 \\ 0 & 0 & -4/21 \end{bmatrix}\begin{bmatrix} 1 & 0 & 0 \\ 0 & 1 & 0 \\ 0 & 0 & -21/4 \end{bmatrix} = \begin{bmatrix} 1 & 0 & 0 \\ 0 & 1 & 0 \\ 0 & 0 & 1 \end{bmatrix}$$

and this makes $A^{-1} = T_7 T_e$:

$$A^{-1} = T_7 T_e$$

$$= \begin{bmatrix} 1 & 0 & 0 \\ 0 & 1 & 0 \\ 0 & 0 & -4/21 \end{bmatrix}\begin{bmatrix} 1/3 & -1/3 & 2/3 \\ 1/21 & 5/21 & -1/21 \\ -1 & 1/4 & 1 \end{bmatrix} = \begin{bmatrix} 1/3 & -1/3 & 2/3 \\ 1/21 & 5/21 & -1/21 \\ 4/21 & -1/21 & -4/21 \end{bmatrix}$$

Thus

$$A^{-1} = \begin{bmatrix} 1/3 & -1/3 & 2/3 \\ 1/21 & 5/21 & -1/21 \\ 4/21 & -1/21 & -4/21 \end{bmatrix}$$

which confirms the result obtained in the example in Section 3.5.

3.10 FINDING THE INVERSE BY THE SWEEP-OUT METHOD

The sweep-out method also involves the use of the elementary transformation matrices that are employed, in turn, to premultiply an augmented matrix consisting of the matrix whose inverse is to be found and the identity matrix. The method has the advantage that the elementary transformation matrices, once used, need not be saved, and the bookkeeping necessary to provide for the cascading of the elementary transformation matrices into an expression for the inverse is not necessary.

The operations on the augmented matrix are continued until the $n \times n$ array at the left is reduced to the identity matrix. At this point the $n \times n$ matrix at the right will be the inverse.

Example: Consider the matrix A:

$$A = \begin{bmatrix} 6 & 3 \\ 2 & 5 \end{bmatrix}$$

Its inverse may be obtained by premultiplying the augmented matrix consisting of A and the identity matrix:

$$A^a = \begin{bmatrix} 6 & 3 & 1 & 0 \\ 2 & 5 & 0 & 1 \end{bmatrix}$$

by selected elementary transformation matrices.

Step One: Change the 6 to a 1 by multiplying by the scaling transformation matrix $T_1 = E_{11}$:

$$T_1 = \begin{bmatrix} 1/6 & 0 \\ 0 & 1 \end{bmatrix}$$

so that $A_1^a = T_1 A^a$

$$A_1^a = \begin{bmatrix} 1/6 & 0 \\ 0 & 1 \end{bmatrix} \begin{bmatrix} 6 & 3 & 1 & 0 \\ 2 & 5 & 0 & 1 \end{bmatrix} = \begin{bmatrix} 1 & 1/2 & 1/6 & 0 \\ 2 & 5 & 0 & 1 \end{bmatrix}$$

Step Two: Multiply row one by -2 and add the result to row two, leaving row one unchanged. This may be done through the use of $T_2 = E_{32}$:

$$T_2 = \begin{bmatrix} 1 & 0 \\ -2 & 1 \end{bmatrix}$$

so that $A_2^a = T_2 A_1^a$:

$$A_2^a = \begin{bmatrix} 1 & 0 \\ -2 & 1 \end{bmatrix} \begin{bmatrix} 1 & 1/2 & 1/6 & 0 \\ 2 & 5 & 0 & 1 \end{bmatrix} = \begin{bmatrix} 1 & 1/2 & 1/6 & 0 \\ 0 & 4 & -1/3 & 1 \end{bmatrix}$$

Step Three: Use the scaling transformation matrix to change the 4 in row two to a 1, $T_3 = E_{13}$:

$$T_3 = \begin{bmatrix} 1 & 0 \\ 0 & 1/4 \end{bmatrix}$$

and this yields $A_3^a = T_3 A_2^a$:

$$A_3^a = \begin{bmatrix} 1 & 0 \\ 0 & 1/4 \end{bmatrix} \begin{bmatrix} 1 & 1/2 & 1/6 & 0 \\ 0 & 4 & -1/3 & 1 \end{bmatrix} = \begin{bmatrix} 1 & 1/2 & 1/6 & 0 \\ 0 & 1 & -1/12 & 1/4 \end{bmatrix}$$

Step Four: Use the combining transformation matrix $T_4 = E_{34}$ to multiply row two by $-1/2$ and add the result to row one, leaving row one unchanged:

$$T_4 = \begin{bmatrix} 1 & -1/2 \\ 0 & 1 \end{bmatrix}$$

so that the final result is $A_4^a = T_4 A_3^a$:

$$A_4^a = \begin{bmatrix} 1 & -1/2 \\ 0 & 1 \end{bmatrix} \begin{bmatrix} 1 & 1/2 & 1/6 & 0 \\ 0 & 1 & -1/12 & 1/4 \end{bmatrix} = \begin{bmatrix} 1 & 0 & 5/24 & -1/8 \\ 0 & 1 & -1/12 & 1/4 \end{bmatrix}$$

The inverse of A is

$$A^{-1} = \begin{bmatrix} 5/24 & -1/8 \\ -1/12 & 1/4 \end{bmatrix}$$

a result that is easily verified by obtaining the determinant of A, swapping the elements on the principal diagonal in A, multiplying the off-diagonal elements of A by -1, and then dividing all four elements by the determinant of A.

3.11 LABOR-SAVING POSSIBILITIES INVOLVING SYMMETRIC MATRICES

Because symmetric matrices are known to have symmetric cofactor matrices and symmetric adjoints that are equal to their cofactor matrices, the evaluation of cofactors may be made less tedious by converting a nonsymmetrical matrix to a symmetrical matrix prior to taking the inverse. Such a procedure eliminates the necessity to evaluate all the cofactors.

The conversion to a symmetrical matrix involves a pre- or postmultiplication by the transpose. Consider the nonsymmetrical matrix A and form B:

$$B = A^T A$$

which is symmetrical. Taking the inverse of B is easier than taking the inverse of A, and after B^{-1} is determined,

$$B^{-1} = (A^T A)^{-1} = A^{-1}(A^T)^{-1}$$

and A^{-1} is reclaimed by a postmultiplication by A^T

$$A^{-1} = B^{-1} A^T = A^{-1}(A^T)^{-1} A^T = A^{-1}$$

If it is desired to form B through a postmultiplication by A^T, then

$$B = AA^T$$
$$B^{-1} = (AA^T)^{-1} = (A^T)^{-1}A^{-1}$$

and this time A^{-1} is reclaimed by a premultiplication by A^T

$$A^{-1} = A^T B^{-1} = A^T(A^T)^{-1}A^{-1} = A^{-1}$$

Example: For the matrix

$$A = \begin{bmatrix} 1 & 2 & 3 \\ 0 & 4 & -1 \\ 1 & 1 & -2 \end{bmatrix}$$

which is nonsymmetric, form $B = A^T A$:

$$B = \begin{bmatrix} 1 & 0 & 1 \\ 2 & 4 & 1 \\ 3 & -1 & -2 \end{bmatrix}\begin{bmatrix} 1 & 2 & 3 \\ 0 & 4 & -1 \\ 1 & 1 & -2 \end{bmatrix} = \begin{bmatrix} 2 & 3 & 1 \\ 3 & 21 & 0 \\ 1 & 0 & 14 \end{bmatrix}$$

which is symmetric.

The determinant of B is

$$\det B = 588 + 0 + 0 - 21 - 0 - 126 = 588 - 147 = 441$$

and B possesses nine cofactors, but only six of them need be evaluated:

$$B_{11} = + \begin{vmatrix} 21 & 0 \\ 0 & 14 \end{vmatrix} = 294$$

$$B_{12} = - \begin{vmatrix} 3 & 0 \\ 1 & 14 \end{vmatrix} = -42$$

$$B_{13} = + \begin{vmatrix} 3 & 21 \\ 1 & 0 \end{vmatrix} = -21$$

$$B_{22} = + \begin{vmatrix} 2 & 1 \\ 1 & 14 \end{vmatrix} = 28 - 1 = 27$$

$$B_{23} = - \begin{vmatrix} 2 & 3 \\ 1 & 0 \end{vmatrix} = -(-3) = 3$$

and

$$B_{33} = + \begin{vmatrix} 2 & 3 \\ 3 & 21 \end{vmatrix} = 42 - 9 = 33$$

Now with $B_{21} = B_{12}$, $B_{31} = B_{13}$, and $B_{32} = B_{23}$,

$$B^C = \text{adj } B = \begin{bmatrix} 294 & -42 & -21 \\ -42 & 27 & 3 \\ -21 & 3 & 33 \end{bmatrix}$$

which are both symmetrical and

$$B^{-1} = (A^T A)^{-1} = \frac{1}{441} \begin{bmatrix} 294 & -42 & -21 \\ -42 & 27 & 3 \\ -21 & 3 & 33 \end{bmatrix}$$

The inverse of A can be obtained from B:

$$A^{-1} = B^{-1} A^T = \frac{1}{441} \begin{bmatrix} 294 & -42 & -21 \\ -42 & 27 & 3 \\ -21 & 3 & 33 \end{bmatrix} \begin{bmatrix} 1 & 0 & 1 \\ 2 & 4 & 1 \\ 3 & -1 & -2 \end{bmatrix}$$

or

$$A^{-1} = \begin{bmatrix} 147/441 & -147/441 & 294/441 \\ 21/441 & 105/441 & -21/441 \\ 84/441 & -21/441 & -84/441 \end{bmatrix} = \begin{bmatrix} 1/3 & -1/3 & 2/3 \\ 1/21 & 5/21 & -1/21 \\ 4/21 & -1/21 & -4/21 \end{bmatrix}$$

which is the result obtained in Sections 3.5 and 3.9.

3.12 EXERCISES

Exercises 3.1 to 3.6 are based on a consideration of the matrices A and B

$$A = \begin{bmatrix} 4 & -2 & -1 \\ 1 & 4 & 0 \\ 1 & 1 & 3 \end{bmatrix} \qquad B = \begin{bmatrix} 6 & -2 & -1 \\ -1 & 4 & 2 \\ 2 & -1 & 4 \end{bmatrix}$$

3.1 Find the inverse of A
3.2 Find the inverse of B
3.3 Find the inverse of AB
3.4 Find the inverse of BA

3.5 Find $A^{-1}B$

3.6 Find $B^{-1}A$

3.7 Find $(A^T)^{-1}$

3.8 Find $(B^T)^{-1}$

3.9 Use successive transformations to find A if

$$A = \begin{bmatrix} 1 & 1 & -1 & -1 \\ 1 & -1 & 1 & 0 \\ 1 & 1 & 0 & 1 \\ 1 & 0 & 1 & -1 \end{bmatrix}$$

3.10 Use the sweep-out method to find A using A in exercise 3.9

3.11 By forming $B = A^T A$, find A^{-1} using A in exercise 3.9

FOUR

PARTITIONING OF MATRICES

4.1 INTRODUCTION

Computations with matrices of higher order can be assisted dramatically if the matrices involved are partitioned, that is, divided into a number of smaller rectangular blocks or submatrices. The partitioning is usually indicated by dashed partitioning lines that must extend entirely through the matrix. For example, the 3×4 matrix M:

$$M = \begin{bmatrix} 1 & -1 & 0 & 2 \\ 0 & 2 & 3 & -1 \\ -1 & 2 & 0 & 4 \end{bmatrix}$$

can be partitioned into

$$M = \left[\begin{array}{cc|cc} 1 & -1 & 0 & 2 \\ 0 & 2 & 3 & -1 \\ \hline -1 & 2 & 0 & 4 \end{array} \right] = \left[\begin{array}{c|c} A & B \\ \hline C & D \end{array} \right]$$

where the matrices A, B, C, and D are given by

$$A = \begin{bmatrix} 1 & -1 \\ 0 & 2 \end{bmatrix} \qquad B = \begin{bmatrix} 0 & 2 \\ 3 & -1 \end{bmatrix}$$

and

$$C = [-1 \quad 2] \qquad D = [0 \quad 4]$$

The arrangement of the partitioned matrices is not unique, as the following examples clearly show:

$$M = \begin{bmatrix} 1 & -1 & 0 & 2 \\ 0 & 2 & 3 & -1 \\ -1 & 2 & 0 & 4 \end{bmatrix} = [A \mid B]$$

$$M = \begin{bmatrix} 1 & -1 & 0 & 2 \\ 0 & 2 & 3 & -1 \\ -1 & 2 & 0 & 4 \end{bmatrix} = \begin{bmatrix} A \\ B \\ C \end{bmatrix}$$

and

$$M = \begin{bmatrix} 1 & -1 & 0 & 2 \\ 0 & 2 & 3 & -1 \\ -1 & 2 & 0 & 4 \end{bmatrix} = \begin{bmatrix} A & B & C \\ D & E & F \end{bmatrix}$$

4.2 ADDITION AND SUBTRACTION OF PARTITIONED MATRICES

Two matrices may be added or subtracted only if they are conformable for these operations, that is, they must be of the the same order. When matrices are partitioned, there is an additional requirement for their addition and subtraction. The matrices must be identically partitioned, that is, the partitioned matrices must be of the same order.

Example: If it is desired to add M that is partitioned into four submatrices A, B, C, and D as indicated:

$$M = \begin{bmatrix} 1 & -1 & 2 \\ 0 & 1 & 1 \\ 1 & 3 & 2 \end{bmatrix} = \begin{bmatrix} A & B \\ C & D \end{bmatrix}$$

to N:

$$N = \begin{bmatrix} 2 & 1 & 1 \\ -1 & 1 & 0 \\ 1 & 2 & 3 \end{bmatrix}$$

then N must also be partitioned into four submatrices E, F, G, and H as indicated:

$$N \begin{bmatrix} 2 & 1 & 1 \\ -1 & 1 & 0 \\ \hline 1 & 2 & 3 \end{bmatrix} = \begin{bmatrix} E & F \\ \hline G & H \end{bmatrix}$$

Observe that corresponding pairs of matrices are of the same order; A and E are 2×2, B and F are 2×1, C and G are 1×2, and D and H are 1×1. The addition may therefore proceed:

$$M + N = \begin{bmatrix} A & B \\ \hline C & D \end{bmatrix} + \begin{bmatrix} E & F \\ \hline G & H \end{bmatrix} = \begin{bmatrix} (A + E) & (B + F) \\ \hline (C + G) & (D + H) \end{bmatrix}$$

and

$$M + N = \begin{bmatrix} \begin{bmatrix} 1 & -1 \\ 0 & 1 \end{bmatrix} + \begin{bmatrix} 2 & 1 \\ -1 & 1 \end{bmatrix} & \begin{bmatrix} 2 \\ 1 \end{bmatrix} + \begin{bmatrix} 1 \\ 0 \end{bmatrix} \\ \hline [1 \quad 3] + [\ 1 \quad 2] & [2] + [3] \end{bmatrix} = \begin{bmatrix} 3 & 0 & 3 \\ -1 & 2 & 1 \\ 2 & 5 & 5 \end{bmatrix}$$

Subtraction, of course, is the inverse of addition, and the foregoing discussion is easily extended to cover subtraction.

4.3 MULTIPLICATION OF PARTITIONED MATRICES

Multiplication of partitioned matrices is accomplished by treating the submatrices derived from the partitioning as single elements. However, in the setup of these elements, careful attention must be paid to the conformability of these elements for their eventual multiplication.

Consider matrix M that is $m \times n$ and that is partitioned into four submatrices A, B, C, and D:

$$M_{m \times n} = \begin{bmatrix} A & B \\ \hline C & D \end{bmatrix}$$

and N that is $p \times q$ and that is partitioned into four submatrices E, F, G, and H:

$$N_{p \times q} = \begin{bmatrix} E & F \\ \hline G & H \end{bmatrix}$$

As discussed in detail in Section 1.7, if $n = p$, M and N are conformable for multiplication, and the product $L = MN$ is $m \times q$.

Now in M, let A be $j \times k$. It is easy to see that B is $j \times (n - k)$; C is $(m - j) \times k$, and D is $(m - j) \times (n - k)$:

$$M_{m \times n} = \left[\begin{array}{c|c} A_{j \times k} & B_{j \times (n-k)} \\ \hline C_{(m-j) \times k} & D_{(m-j) \times (n-k)} \end{array} \right]$$

The question then becomes: how should E, F, G, and H in N be arranged so that the product

$$L = MN = \left[\begin{array}{c|c} A & B \\ \hline C & D \end{array} \right] \left[\begin{array}{c|c} E & F \\ \hline G & H \end{array} \right] = \left[\begin{array}{c|c} (AE + BG) & (AF + BH) \\ \hline (CE + DG) & (CF + DH) \end{array} \right]$$

can be executed?

Let E by $r \times s$ so that F is $r \times (q - s)$, G is $(p - r) \times s$ and H is $(p - r) \times (q - s)$. If A which is $j \times k$ is to premultiply E, then $r = k$, and the product AE, as well as the product BG, will be $j \times s$. Moreover, in the product BG where B is $j \times (n - k)$ and G is $(p - r) \times s$, conformability dictates that $n - k = p - r$. This, of course, means that if $r = k$, then $p = n$ which is required if the product $L = MN$ is to be executed at all. These are, however, the only requirements

$$p = n$$

$$r = k$$

Example: It is desired to use partitioning obtain the product of the partitioned matrix M which is 4×5

$$M = \left[\begin{array}{ccc|cc} 1 & 0 & 2 & 3 & -1 \\ 0 & 1 & 1 & -1 & 2 \\ 3 & 2 & 1 & 2 & 3 \\ \hline -1 & 1 & 1 & 0 & 2 \end{array} \right] = \left[\begin{array}{c|c} A & B \\ \hline C & D \end{array} \right]$$

with the matrix N which is 5×5:

$$N = \left[\begin{array}{ccccc} 4 & -1 & -1 & 0 & 0 \\ -1 & 3 & 0 & 0 & -1 \\ -1 & 0 & 4 & -1 & -1 \\ 0 & 0 & -1 & 3 & -1 \\ 0 & -1 & -1 & -1 & 4 \end{array} \right] = \left[\begin{array}{c|c} E & F \\ \hline G & H \end{array} \right]$$

The first step in finding the product $L = MN$ involves the proper selection of the order of E. Observe first that $M = 4 \times 5$ and $N = 5 \times 5$. Thus $n = 5 = p = 5$, and M may premultiply N. However, A is $3 \times$

3 ($j = k = 3$) so that E, which is $r \times s$ must have $r = k = 3$. The value of s is discretionary; let it be set at $s = 2$. Then

$$N = \left[\begin{array}{cc:ccc} 4 & -1 & -1 & 0 & 0 \\ -1 & 3 & 0 & 0 & -1 \\ -1 & 0 & 4 & -1 & -1 \\ \hdashline 0 & 0 & -1 & 3 & -1 \\ 0 & -1 & -1 & -1 & 4 \end{array}\right] = \left[\begin{array}{c:c} E & F \\ \hdashline G & H \end{array}\right]$$

The product $L = MN$ will be

$$L = \left[\begin{array}{c:c} A & B \\ \hdashline C & D \end{array}\right]\left[\begin{array}{c:c} E & F \\ \hdashline G & H \end{array}\right] = \left[\begin{array}{c:c} T & U \\ \hdashline V & W \end{array}\right] = \left[\begin{array}{c:c} (AE + BG) & (AF + BH) \\ \hdashline (CE + DG) & (CF + DH) \end{array}\right]$$

with

$$T = AE + BG = \begin{bmatrix} 1 & 0 & 2 \\ 0 & 1 & 1 \\ 3 & 2 & 1 \end{bmatrix}\begin{bmatrix} 4 & -1 \\ -1 & 3 \\ -1 & 0 \end{bmatrix} + \begin{bmatrix} 3 & -1 \\ -1 & 2 \\ 2 & 3 \end{bmatrix}\begin{bmatrix} 0 & 0 \\ 0 & -1 \end{bmatrix}$$

or

$$T = \begin{bmatrix} 2 & -1 \\ -2 & 3 \\ 9 & 3 \end{bmatrix} + \begin{bmatrix} 0 & 1 \\ 0 & -2 \\ 0 & -3 \end{bmatrix} = \begin{bmatrix} 2 & 0 \\ -2 & 1 \\ 9 & 0 \end{bmatrix}$$

Next

$$U = AF + BH$$

$$= \begin{bmatrix} 1 & 0 & 2 \\ 0 & 1 & 1 \\ 3 & 2 & 1 \end{bmatrix}\begin{bmatrix} -1 & 0 & 0 \\ 0 & 0 & -1 \\ 4 & -1 & -1 \end{bmatrix} + \begin{bmatrix} 3 & -1 \\ -1 & 2 \\ 2 & 3 \end{bmatrix}\begin{bmatrix} -1 & 3 & -1 \\ -1 & -1 & 4 \end{bmatrix}$$

or

$$U = \begin{bmatrix} 7 & -2 & -2 \\ 4 & -1 & -2 \\ 1 & -1 & -3 \end{bmatrix} + \begin{bmatrix} -2 & 10 & -7 \\ -1 & -5 & 9 \\ -5 & 3 & 10 \end{bmatrix} = \begin{bmatrix} 5 & 8 & -9 \\ 3 & -6 & 7 \\ -4 & 2 & 7 \end{bmatrix}$$

Then

$$V = CE + DG = \begin{bmatrix} -1 & 1 & 1 \end{bmatrix} \begin{bmatrix} -4 & -1 \\ -1 & 3 \\ -1 & 0 \end{bmatrix} + \begin{bmatrix} 0 & 2 \end{bmatrix} \begin{bmatrix} 0 & 0 \\ 0 & -1 \end{bmatrix}$$

or

$$V = \begin{bmatrix} -6 & 4 \end{bmatrix} + \begin{bmatrix} 0 & -2 \end{bmatrix} = \begin{bmatrix} -6 & 2 \end{bmatrix}$$

Finally,

$$W = CF + DH$$

$$= \begin{bmatrix} -1 & 1 & 1 \end{bmatrix} \begin{bmatrix} -1 & 0 & 0 \\ 0 & 0 & -1 \\ 4 & -1 & -1 \end{bmatrix} + \begin{bmatrix} 0 & 2 \end{bmatrix} \begin{bmatrix} -1 & 3 & -1 \\ -1 & -1 & 4 \end{bmatrix}$$

or

$$W = \begin{bmatrix} 5 & -1 & -2 \end{bmatrix} + \begin{bmatrix} -2 & -2 & 8 \end{bmatrix} = \begin{bmatrix} 3 & -3 & 6 \end{bmatrix}$$

The result $L = MN$ is

$$L = \begin{bmatrix} 2 & 0 & 5 & 8 & -9 \\ -2 & 1 & 3 & -6 & 7 \\ 9 & 0 & -4 & 2 & 7 \\ -6 & 2 & 3 & -3 & 6 \end{bmatrix}$$

4.4 TRANSPOSE OF A PARTITIONED MATRIX

It is easy to verify that the transpose of a partitioned matrix such as

$$M = \begin{bmatrix} A & B \\ C & D \end{bmatrix}$$

is

$$M^T = \begin{bmatrix} A^T & C^T \\ B^T & D^T \end{bmatrix}$$

4.5 INVERSE OF A NONSYMMETRICAL MATRIX BY PARTITIONING

Consider a matrix M and let it be partitioned into four submatrices A, B, C, and D such that A is square and nonsingular:

$$M = \left[\begin{array}{c|c} A & B \\ \hline C & D \end{array}\right]$$

Assume that M has an inverse so that the system of simultaneous, linear algebraic equations

$$MX = Y$$

can be solved by premultiplying both sides by M^{-1}:

$$M^{-1}MX = X = M^{-1}Y$$

in which M^{-1} is the partitioned matrix:

$$M = \left[\begin{array}{c|c} E & F \\ \hline G & H \end{array}\right]$$

The representation for $Y = MX$ where X and Y must also be partitioned is

$$Y = MX = \left[\begin{array}{c|c} A & B \\ \hline C & D \end{array}\right]\left[\begin{array}{c} X_1 \\ \hline X_2 \end{array}\right] = \left[\begin{array}{c} Y_1 \\ \hline Y_2 \end{array}\right]$$

and this may be expanded into the pair of matric equations

$$AX_1 + BX_2 = Y_1 \tag{4.1a}$$

and

$$CX_1 + DX_2 = Y_2 \tag{4.1b}$$

But X may be determined from Y by employing the inverse of M:

$$X = M^{-1}Y = \left[\begin{array}{c|c} E & F \\ \hline G & H \end{array}\right]\left[\begin{array}{c} Y_1 \\ \hline Y_2 \end{array}\right] = \left[\begin{array}{c} X_1 \\ \hline X_2 \end{array}\right]$$

and these, too, may be expanded into a pair of matric equations:

$$EY_1 + FY_2 = X_1 \tag{4.2a}$$

and

$$GY_1 + HY_2 = X_2 \tag{4.2b}$$

Expressions for E, F, G, and H may be developed from Eqs. (4.1).

Because A is required to be square and nonsingular, Eq. (4.1a) can be solved for X_1:

$$X_1 = A^{-1}Y_1 - A^{-1}BX_2 \qquad (4.3)$$

and if this is put into Eq. (4.1b) so that

$$C(A^{-1}Y_1 - A^{-1}BX_2) + DX_2 = Y_2$$

then X_2 may be determined:

$$(D - CA^{-1}B)X_2 = Y_2 - CA^{-1}Y_1$$

or

$$X_2 = (D - CA^{-1}B)^{-1}Y_2 - (D - CA^{-1}B)^{-1}CA^{-1}Y_1 \qquad (4.4)$$

Let

$$P = A^{-1}B \qquad (4.5)$$

and

$$Q = D - CA^{-1}B = D - CP \qquad (4.6)$$

With P and Q so defined, and noting that Q must be nonsingular, Eq. (4.3) may be adjusted to

$$X_1 = A^{-1}Y_1 - PX_2 \qquad (4.7)$$

and so may Eq. (4.4):

$$X_2 = -Q^{-1}CA^{-1}Y_1 + Q^{-1}Y_2 \qquad (4.8)$$

Now if Eq. (4.8) is put into Eq. (4.7), the result is

$$X_1 = A^{-1}Y_1 - P(-Q^{-1}CA^{-1}Y_1 + Q^{-1}Y_2)$$

or

$$X_1 = (A^{-1} + PQ^{-1}CA^{-1})Y_1 - PQ^{-1}Y_2 \qquad (4.9)$$

Relationships for the submatrices E, F, G, and H of the inverse of M can now be obtained from a direct comparison of Eqs. (4.2), (4.8), and (4.9):

$$E = A^{-1} + PQ^{-1}CA^{-1} \qquad (4.10a)$$

$$F = -PQ^{-1} = -A^{-1}BQ^{-1} \qquad (4.10b)$$

$$G = -Q^{-1}CA^{-1} \qquad (4.10c)$$

and

$$H = Q^{-1} \qquad (4.10d)$$

These may be summarized by displaying them in the partitioned matrix

$$M^{-1} = \left[\begin{array}{c|c} E & F \\ \hline G & H \end{array}\right] = \left[\begin{array}{c|c} A^{-1} + PQ^{-1}CA^{-1} & -PQ^{-1} \\ \hline -Q^{-1}CA^{-1} & Q^{-1} \end{array}\right] \quad (4.11)$$

Example: Consider the problem of finding the inverse of the matrix M by partitioning

$$M = \left[\begin{array}{cccc} 4 & -1 & 0 & -1 \\ -2 & 3 & 0 & 0 \\ 1 & 1 & 4 & 1 \\ 1 & 0 & -1 & 3 \end{array}\right]$$

The method available consists of recognizing that M is to be partitioned as

$$M = \left[\begin{array}{cc} A & B \\ C & D \end{array}\right]$$

where A must be square and nonsingular. In this case, the inverse is also a partitioned matrix:

$$M^{-1} = \left[\begin{array}{cc} E & F \\ G & H \end{array}\right]$$

where the submatrices E, F, G, and H may be determined from Eqs. (4.10) or (4.11).

Here A may be chosen as 1×1, 2×2, or 3×3. Select

$$A = \left[\begin{array}{cc} 4 & -1 \\ -2 & 3 \end{array}\right]$$

so that A^{-1} is quickly obtained, because $\det A = 12 - 2 = 10$:

$$A^{-1} = \left[\begin{array}{cc} 3/10 & 1/10 \\ 1/5 & 2/5 \end{array}\right]$$

and

$$B = \left[\begin{array}{cc} 0 & -1 \\ 0 & 0 \end{array}\right]$$

$$C = \left[\begin{array}{cc} 1 & 1 \\ 1 & 0 \end{array}\right]$$

and

$$D = \begin{bmatrix} 4 & 1 \\ -1 & 3 \end{bmatrix}$$

This makes P, by Eq. (4.5),

$$P = A^{-1}B = \begin{bmatrix} 3/10 & 1/10 \\ 1/5 & 2/5 \end{bmatrix} \begin{bmatrix} 0 & -1 \\ 0 & 0 \end{bmatrix} = \begin{bmatrix} 0 & -3/10 \\ 0 & -1/5 \end{bmatrix}$$

and Q, by Eq. (4.6),

$$Q = D - CP = \begin{bmatrix} 4 & 1 \\ -1 & 3 \end{bmatrix} - \begin{bmatrix} 1 & 1 \\ 1 & 0 \end{bmatrix} \begin{bmatrix} 0 & -3/10 \\ 0 & -1/5 \end{bmatrix}$$

$$= \begin{bmatrix} 4 & 1 \\ -1 & 3 \end{bmatrix} - \begin{bmatrix} 0 & -1/2 \\ 0 & -3/10 \end{bmatrix}$$

or

$$Q = \begin{bmatrix} 4 & 3/2 \\ -1 & 33/10 \end{bmatrix} = \begin{bmatrix} 40/10 & 15/10 \\ -10/10 & 33/10 \end{bmatrix}$$

The determinant of Q is

$$\det Q = \frac{1}{100}(1320 + 150) = \frac{1470}{100} = \frac{147}{10}$$

and

$$Q^{-1} = \frac{10}{147} \begin{bmatrix} 33/10 & -15/10 \\ 10/10 & 40/10 \end{bmatrix} = \begin{bmatrix} 33/147 & -15/147 \\ 10/147 & 40/147 \end{bmatrix}$$

Then

$$PQ^{-1} = \frac{1}{147} \begin{bmatrix} 0 & -3/10 \\ 0 & -1/5 \end{bmatrix} \begin{bmatrix} 33 & -15 \\ 10 & 40 \end{bmatrix} = \frac{1}{147} \begin{bmatrix} -3 & -12 \\ -2 & -8 \end{bmatrix}$$

$$CA^{-1} = \begin{bmatrix} 1 & 1 \\ 1 & 0 \end{bmatrix} \begin{bmatrix} 3/10 & 1/10 \\ 1/5 & 2/5 \end{bmatrix} = \begin{bmatrix} 1/2 & 1/2 \\ 3/10 & 1/10 \end{bmatrix}$$

and

$$PQ^{-1}CA^{-1} = \frac{1}{147} \frac{1}{10} \begin{bmatrix} -3 & -12 \\ -2 & -8 \end{bmatrix} \begin{bmatrix} 5 & 5 \\ 3 & 1 \end{bmatrix} = \frac{1}{1470} \begin{bmatrix} -51 & -27 \\ -34 & -18 \end{bmatrix}$$

Now E can be established from Eq. (4.10a):

$$E = A^{-1} + PQ^{-1}CA^{-1} = \begin{bmatrix} 3/10 & 1/10 \\ 1/5 & 2/5 \end{bmatrix} + \frac{1}{1470}\begin{bmatrix} -51 & -27 \\ -34 & -18 \end{bmatrix}$$

$$= \frac{1}{1470}\left[\begin{bmatrix} 441 & 147 \\ 294 & 588 \end{bmatrix} + \begin{bmatrix} -51 & -27 \\ -34 & -18 \end{bmatrix}\right]$$

or

$$E = \frac{1}{1470}\begin{bmatrix} 390 & 120 \\ 260 & 570 \end{bmatrix} = \begin{bmatrix} 39/147 & 12/147 \\ 26/147 & 57/147 \end{bmatrix}$$

Next F is determined. First,

$$BQ^{-1} = \frac{1}{147}\begin{bmatrix} 0 & -1 \\ 0 & 0 \end{bmatrix}\begin{bmatrix} 33 & -15 \\ 10 & 40 \end{bmatrix} = \frac{1}{147}\begin{bmatrix} -10 & -40 \\ 0 & 0 \end{bmatrix}$$

and then by Eq. (4.10b):

$$F = -A^{-1}BQ^{-1}$$

$$= -\frac{1}{147}\frac{1}{10}\begin{bmatrix} 3 & 1 \\ 2 & 4 \end{bmatrix}\begin{bmatrix} -10 & -40 \\ 0 & 0 \end{bmatrix} = -\frac{1}{1470}\begin{bmatrix} -30 & -120 \\ -20 & -80 \end{bmatrix}$$

or

$$F = \begin{bmatrix} 3/147 & 12/147 \\ 2/147 & 8/147 \end{bmatrix}$$

The submatrix G is obtained from Eq. (4.10c):

$$G = -Q^{-1}CA^{-1}$$

$$= -\frac{1}{147}\frac{1}{10}\begin{bmatrix} 33 & -15 \\ 10 & 40 \end{bmatrix}\begin{bmatrix} 5 & 5 \\ 3 & 1 \end{bmatrix} = -\frac{1}{1470}\begin{bmatrix} 120 & 150 \\ 170 & 90 \end{bmatrix}$$

or

$$G = \begin{bmatrix} -12/147 & -15/147 \\ -17/147 & -9/147 \end{bmatrix}$$

The value of H is nothing more than Q^{-1}, as shown by Eq. (4.10d):

$$H = Q^{-1} = \begin{bmatrix} 33/147 & -15/147 \\ 10/147 & 40/147 \end{bmatrix}$$

The inverse of M is

$$M^{-1} = \begin{bmatrix} 39/147 & 12/147 & 3/147 & 12/147 \\ 26/147 & 57/147 & 2/147 & 8/147 \\ -12/147 & -15/147 & 33/147 & -15/147 \\ -17/147 & -9/147 & 10/147 & 40/147 \end{bmatrix}$$

The requirement that $MM^{-1} = I$ should be checked, and the reader may wish to verify that

$$\begin{bmatrix} 39 & 12 & 3 & 12 \\ 26 & 57 & 2 & 8 \\ -12 & -15 & 33 & -15 \\ -17 & -9 & 10 & 40 \end{bmatrix}\begin{bmatrix} 4 & -1 & 0 & -1 \\ -2 & 3 & 0 & 0 \\ 1 & 1 & 4 & 1 \\ 1 & 0 & -1 & 3 \end{bmatrix}$$

$$= \begin{bmatrix} 147 & 0 & 0 & 0 \\ 0 & 147 & 0 & 0 \\ 0 & 0 & 147 & 0 \\ 0 & 0 & 0 & 147 \end{bmatrix}$$

4.6 INVERSE OF A SYMMETRICAL MATRIX BY PARTITIONING

If the matrix

$$M = \begin{bmatrix} A & B \\ C & D \end{bmatrix}$$

is symmetric, then the submatrices E and G in its inverse

$$M^{-1} = \begin{bmatrix} E & F \\ G & H \end{bmatrix}$$

as given by Eqs. (4.10a) and (4.10c) may be simplified to some extent.

In the symmetrical matrix,

$$B = C^{\mathrm{T}}$$

$$C = B^{\mathrm{T}}$$

and A, which must be square and nonsingular, has a symmetrical inverse that is equal to its transpose:

$$A^{-1} = (A^{-1})^{\mathrm{T}}$$

Under these circumstances, P, which is given by Eq. (4.5),

$$P = A^{-1}B$$

has a transpose

$$P^T = (A^{-1}B)^T = B^T(A^{-1})^T = CA^{-1}$$

so that E as given by Eq. (4.10a)

$$E = A^{-1} + PQ^{-1}CA^{-1}$$

can be written as

$$E = A^{-1} + PQ^{-1}P^T \tag{4.12a}$$

and G as given by Eq. (4.10c)

$$G = -Q^{-1}CA^{-1}$$

can be written as

$$G = -Q^{-1}P^T = -(PQ^{-1})^T \tag{4.12b}$$

For the case of matrices that are symmetric, the inverse can then be summarized as

$$M = \begin{bmatrix} E & F \\ \hline G & H \end{bmatrix} = \left[\begin{array}{c|c} A^{-1} + PQ^{-1}P^T & -PQ^{-1} \\ \hline -Q^{-1}P^T = -(PQ^{-1})^T & Q^{-1} \end{array} \right] \tag{4.13}$$

Example: Consider the problem of finding the inverse of the matrix M by partitioning

$$M = \begin{bmatrix} 4 & -1 & -1 & -1 & 0 \\ -1 & 6 & -2 & -1 & -1 \\ -1 & -2 & 8 & -2 & 0 \\ -1 & -1 & -2 & 6 & 0 \\ 0 & -1 & 0 & 0 & 2 \end{bmatrix}$$

The method available consists of recognizing that M is symmetric and that it can be partitioned into four submatrices:

$$M = \begin{bmatrix} A & B \\ C & D \end{bmatrix}$$

In this case the inverse is also a partitioned matrix:

$$M^{-1} = \begin{bmatrix} E & F \\ G & H \end{bmatrix}$$

where the submatrices E, F, G, and H can be determined from Eqs. (4.12a), (4.10b), (4.12b), (4.10d), or (4.13).

Here A must be chosen square as 2×2, 3×3, or 4×4. Let it be 3×3:

$$A = \begin{bmatrix} 4 & -1 & -1 \\ -1 & 6 & -2 \\ -1 & -2 & 8 \end{bmatrix}$$

and for this symmetrical matrix:

$$\det A = 192 - 2 - 2 - 6 - 16 - 8 = 158$$

Then because the cofactor matrix is equal, in this case, to the adjoint

$$\text{adj } A = \begin{bmatrix} + \begin{vmatrix} 6 & -2 \\ -2 & 8 \end{vmatrix} & - \begin{vmatrix} -1 & -2 \\ -1 & 8 \end{vmatrix} & + \begin{vmatrix} -1 & 6 \\ -1 & -2 \end{vmatrix} \\ - \begin{vmatrix} -1 & -1 \\ -2 & 8 \end{vmatrix} & + \begin{vmatrix} 4 & -1 \\ -1 & 8 \end{vmatrix} & - \begin{vmatrix} 4 & -1 \\ -1 & -2 \end{vmatrix} \\ + \begin{vmatrix} -1 & -1 \\ 6 & -2 \end{vmatrix} & - \begin{vmatrix} 4 & -1 \\ -1 & -2 \end{vmatrix} & + \begin{vmatrix} 4 & -1 \\ -1 & 6 \end{vmatrix} \end{bmatrix}$$

or

$$\text{adj } A = \begin{bmatrix} 44 & 10 & 8 \\ 10 & 31 & 9 \\ 8 & 9 & 23 \end{bmatrix}$$

the inverse of A is

$$A^{-1} = \frac{\text{adj } A}{\det A} = \frac{1}{158} \begin{bmatrix} 44 & 10 & 8 \\ 10 & 31 & 9 \\ 8 & 9 & 23 \end{bmatrix}$$

and the submatrices B, C, and D of M are

$$B = \begin{bmatrix} -1 & 0 \\ -1 & -1 \\ -2 & 0 \end{bmatrix}$$

$$C = B^{\mathrm{T}} = \begin{bmatrix} -1 & -1 & -2 \\ 0 & -1 & 0 \end{bmatrix}$$

and

$$D = \begin{bmatrix} 6 & 0 \\ 0 & 2 \end{bmatrix}$$

The matrix P is evaluated from Eq. (4.5):

$$P = A^{-1}B = \frac{1}{158} \begin{bmatrix} 44 & 10 & 8 \\ 10 & 31 & 9 \\ 8 & 9 & 23 \end{bmatrix} \begin{bmatrix} -1 & 0 \\ -1 & -1 \\ -2 & 0 \end{bmatrix} = \frac{1}{158} \begin{bmatrix} -70 & -10 \\ -59 & -31 \\ -63 & -9 \end{bmatrix}$$

and then Q is obtained from Eq. (4.6):

$$Q = D - CP = \begin{bmatrix} 6 & 0 \\ 0 & 2 \end{bmatrix} - \begin{bmatrix} -1 & -1 & -2 \\ 0 & -1 & 0 \end{bmatrix} \frac{1}{158} \begin{bmatrix} -70 & -10 \\ -59 & -31 \\ -63 & -9 \end{bmatrix}$$

$$= \begin{bmatrix} 6 & 0 \\ 0 & 2 \end{bmatrix} - \frac{1}{158} \begin{bmatrix} 255 & 59 \\ 59 & 31 \end{bmatrix}$$

and, finally,

$$Q = \frac{1}{158} \begin{bmatrix} 693 & -59 \\ -59 & 285 \end{bmatrix}$$

The inverse of Q will be needed. First observe that Q is also symmetric and compute det Q

$$\det Q = \frac{693(285) - 59^2}{158^2} = \frac{1228}{158}$$

and

$$Q^{-1} = \frac{158}{1228} \begin{bmatrix} 285/158 & 59/158 \\ 59/158 & 693/158 \end{bmatrix} = \frac{1}{1228} \begin{bmatrix} 285 & 59 \\ 59 & 693 \end{bmatrix}$$

The matrix P has a transpose

$$P^T = \frac{1}{158} \begin{bmatrix} -70 & -59 & -63 \\ -10 & -31 & -9 \end{bmatrix}$$

and all the matrices needed to evaluate E using Eq. (4.12a) are now available. First,

$$Q^{-1}P^T = \frac{1}{1228} \begin{bmatrix} 285 & 59 \\ 59 & 693 \end{bmatrix} \frac{1}{158} \begin{bmatrix} -70 & -59 & -63 \\ -10 & -31 & -9 \end{bmatrix}$$

$$= -\frac{1}{1228} \begin{bmatrix} 130 & 118 & 117 \\ 70 & 158 & 63 \end{bmatrix}$$

and then

$$PQ^{-1}P^{\mathrm{T}} = -\frac{1}{158}\begin{bmatrix} -70 & -10 \\ -59 & -31 \\ -63 & -9 \end{bmatrix}\frac{1}{1228}\begin{bmatrix} 130 & 118 & 117 \\ 70 & 158 & 63 \end{bmatrix}$$

or

$$PQ^{-1}P^{\mathrm{T}} = \frac{1}{158(1228)}\begin{bmatrix} 9800 & 9840 & 8820 \\ 9840 & 11{,}860 & 8856 \\ 8820 & 8856 & 7938 \end{bmatrix}$$

Now go for E by Eq. (4.12a), $E = A^{-1} + PQ^{-1}P^{\mathrm{T}}$:

$$E = \frac{1}{158}\begin{bmatrix} 44 & 10 & 8 \\ 10 & 31 & 9 \\ 8 & 9 & 23 \end{bmatrix} + \frac{1}{158(1228)}\begin{bmatrix} 9800 & 9840 & 8820 \\ 9840 & 11{,}860 & 8856 \\ 8820 & 8856 & 7938 \end{bmatrix}$$

and the reader may wish to verify that

$$E = \frac{1}{1228}\begin{bmatrix} 404 & 140 & 118 \\ 140 & 316 & 126 \\ 118 & 126 & 229 \end{bmatrix}$$

The next step is to determine F, and there is a bonus here, because the matrix M is symmetric and $F = G^{\mathrm{T}}$. From Eq. (4.10b),

$$F = G^{\mathrm{T}} = -PQ^{-1} = -\frac{1}{158}\begin{bmatrix} -70 & -10 \\ -59 & -31 \\ -63 & -9 \end{bmatrix}\frac{1}{1228}\begin{bmatrix} 285 & 59 \\ 59 & 693 \end{bmatrix}$$

$$= \frac{1}{158(1228)}\begin{bmatrix} 20{,}540 & 11{,}060 \\ 18{,}644 & 24{,}694 \\ 18{,}486 & 9954 \end{bmatrix}$$

or

$$F = G^{\mathrm{T}} = \frac{1}{1228}\begin{bmatrix} 130 & 70 \\ 118 & 158 \\ 117 & 63 \end{bmatrix}$$

Finally, in accordance with Eq. (4.10d), H is nothing more than Q^{-1}:

$$H = Q^{-1} = \frac{1}{1228}\begin{bmatrix} 285 & 59 \\ 59 & 693 \end{bmatrix}$$

and when the foregoing determinations of E, F, G, and H are all put together, the value of M^{-1} is seen to be

$$M^{-1} = \frac{1}{1228} \begin{bmatrix} 404 & 140 & 118 & 130 & 70 \\ 140 & 316 & 126 & 118 & 158 \\ 118 & 126 & 229 & 117 & 63 \\ 130 & 118 & 117 & 285 & 59 \\ 70 & 158 & 63 & 59 & 693 \end{bmatrix}$$

and the reader may wish to verify that the matrix product

$$\begin{bmatrix} 404 & 140 & 118 & 130 & 70 \\ 140 & 316 & 126 & 118 & 158 \\ 118 & 126 & 229 & 117 & 63 \\ 130 & 118 & 117 & 285 & 59 \\ 70 & 158 & 63 & 59 & 693 \end{bmatrix} \begin{bmatrix} 4 & -1 & -1 & -1 & 0 \\ -1 & 6 & -2 & -1 & -1 \\ -1 & -2 & 8 & -2 & 0 \\ -1 & -1 & -2 & 6 & 0 \\ 0 & -1 & 0 & 0 & 2 \end{bmatrix}$$

does indeed equal the product of the scalar, 1228, and the identity matrix:

$$1228 \begin{bmatrix} 1 & 0 & 0 & 0 & 0 \\ 0 & 1 & 0 & 0 & 0 \\ 0 & 0 & 1 & 0 & 0 \\ 0 & 0 & 0 & 1 & 0 \\ 0 & 0 & 0 & 0 & 1 \end{bmatrix}$$

4.7 EXERCISES

Exercises 4.1 through 4.4 are based on a consideration of the matrices A and B:

$$A = \begin{bmatrix} 1 & 2 & 2 & 2 & -1 \\ 1 & 0 & -1 & 2 & 0 \\ 0 & -2 & -2 & 1 & 1 \\ 0 & 0 & -1 & -2 & 1 \\ 2 & 2 & 0 & 0 & -1 \\ 1 & -1 & 1 & 0 & 0 \end{bmatrix}$$

and

$$B = \begin{bmatrix} 1 & 0 & 0 & -1 & 2 \\ 1 & 2 & -1 & 2 & 1 \\ 1 & 0 & -1 & 1 & 0 \\ 2 & -1 & 0 & 0 & 0 \\ 1 & 1 & 1 & 0 & 0 \end{bmatrix}$$

4.1 Find the product AB without resorting to partitioning.

4.2 Find the product AB by making a 2×2 partition of matrix A with the matrix at the upper left 2×2.

4.3 Repeat exercise 4.2 with the matrix at the upper left of A 3×3.

4.4 Repeat exercise 4.2 with the matrix at the upper left of A 4×4.

4.5 Find M^{-1} via partitioning if

$$M = \begin{bmatrix} A & B \\ C & D \end{bmatrix} = \begin{bmatrix} 1 & 1 & -1 & -1 \\ 1 & -1 & 1 & 0 \\ 1 & 1 & 0 & 1 \\ 1 & 0 & 1 & -1 \end{bmatrix}$$

4.6 Find M^{-1} via partitioning if

$$M = \begin{bmatrix} A & B \\ C & D \end{bmatrix} = \begin{bmatrix} 4 & 1 & 1 & -1 \\ 1 & 3 & -2 & 0 \\ 1 & -2 & 3 & 0 \\ -1 & 0 & 0 & 3 \end{bmatrix}$$

4.7 Find M^{-1} via partitioning if

$$M = \begin{bmatrix} A & B \\ C & D \end{bmatrix} = \begin{bmatrix} 2 & -1 & -1 & 0 & 0 \\ -1 & 1 & 0 & -1 & 0 \\ -1 & 0 & 1 & 1 & 1 \\ 0 & -1 & 1 & 1 & -1 \\ 0 & 0 & 1 & -1 & 1 \end{bmatrix}$$

FIVE

SIMULTANEOUS EQUATIONS

5.1 INTRODUCTION

Systems of simultaneous, linear algebraic equations occur quite often in engineering analysis in all the engineering disciplines. It has been observed in previous chapters that these systems of equations can be represented by the matric shorthand

$$AX = B$$

where A is a coefficient matrix, X is a column vector of solutions, and B is a column vector of constants. If the system is "driven" by a set of variables, then the representation of the system of equations becomes

$$AX = Y$$

where Y is a column vector of the driving or forcing variables.

Many methods of solution exist. For example, a considerable amount of space in this book has been devoted to a discussion of the inverse of a matrix. If the inverse of A exists, that is, if A is not singular, a solution for X has been observed to be

$$X = A^{-1}Y$$

Of course, this study has indicated under what conditions the matrix A becomes singular. But what about the existence of a solution to $AX = B$, and are there better methods of finding a solution? The answers to questions such as these and others is the subject of this chapter.

5.2 LINEAR DEPENDENCE AND INDEPENDENCE

The coefficients in a set of simultaneous, linear algebraic equations can be represented as a set of row or column vectors; that is, in $AX = B$, the coefficient matrix A can be represented as

$$A = [A_1 \quad A_2 \quad A_3 \quad \cdots \quad A_n]$$

or as

$$A = \begin{bmatrix} A_1 \\ A_2 \\ A_3 \\ \cdots \\ A_n \end{bmatrix}$$

Observe that all these vectors have the same order; for the row vectors, they are $1 \times n$, and for the column vectors, they are $n \times 1$.

In either case the set of vectors are said to be a *linearly dependent set* if in

$$\alpha_1 A_1 + \alpha_2 A_2 + \alpha_3 A_3 + \cdots + \alpha_n A_n = 0 \qquad (5.1)$$

there exists at least one value of the scalars α that is not identically equal to zero. If all α are equal to zero, the set of vectors is said to be *linearly independent*, and the vectors then constitute a linearly independent set. Observe that if a set of vectors is not linearly independent, that set must be linearly dependent.

Example: Consider the matrix

$$A = \begin{bmatrix} 3 & -2 & 0 \\ -2 & 4 & -1 \\ 0 & -1 & 6 \end{bmatrix}$$

which was shown in Section 3.5 to possess an inverse. This matrix may be represented as three row vectors:

$$A_1 = [\ 3 \quad -2 \quad 0]$$
$$A_2 = [-2 \quad 4 \quad -1]$$

and

$$A_3 = [\ 0 \quad -1 \quad 6]$$

Equation (5.1) is used to determine whether these three vectors constitute a linearly independent set:

$$\alpha_1[3 \quad -2 \quad 0] + \alpha_2[-2 \quad 4 \quad -1] + \alpha_3[0 \quad -1 \quad 6] = [0 \quad 0 \quad 0]$$

Expansion leads to the set of three equations in three unknowns:

$$3\alpha_1 - 2\alpha_2 = 0$$

$$-2\alpha_1 + 4\alpha_2 - \alpha_3 = 0$$

$$-\alpha_2 + 6\alpha_3 = 0$$

and it is noted that from the third of these, $\alpha_3 = \alpha_2/6$, and from the first, $\alpha_1 = 2\alpha_2/3$. With these in the second equation

$$-2(2\alpha_2/3) + 4\alpha_2 - (\alpha_2/6) = 0$$

it is seen that $\alpha_2 = 0$ and that $\alpha_1 = \alpha_2 = \alpha_3 = 0$. These mean that the set of row vectors form a linearly independent set.

Example: The matrix

$$A = \begin{bmatrix} 4 & -2 & 0 \\ 2 & -1 & 0 \\ 0 & 2 & 3 \end{bmatrix}$$

may be represented by a set of column vectors

$$A_1 = \begin{bmatrix} 4 \\ 2 \\ 0 \end{bmatrix} \quad A_2 = \begin{bmatrix} -2 \\ -1 \\ 2 \end{bmatrix} \quad \text{and} \quad A_3 = \begin{bmatrix} 0 \\ 0 \\ 3 \end{bmatrix}$$

Equation (5.1) may also be used in this case to determine whether these column vectors form a linearly independent or linearly dependent set:

$$\alpha_1 \begin{bmatrix} 4 \\ 2 \\ 0 \end{bmatrix} + \alpha_2 \begin{bmatrix} -2 \\ -1 \\ 2 \end{bmatrix} + \alpha_3 \begin{bmatrix} 0 \\ 0 \\ 3 \end{bmatrix} = \begin{bmatrix} 0 \\ 0 \\ 0 \end{bmatrix}$$

The check for linear independence is based on an investigation of the values of the α's in the expanded version:

$$4\alpha_1 - 2\alpha_2 = 0$$

$$2\alpha_1 - \alpha_2 = 0$$

$$+2\alpha_2 + 3\alpha_3 = 0$$

From the first and second of these equations, $\alpha_2 = 2\alpha_1$. This means that the third equation can be expressed in terms of either of the constants

$$3\alpha_3 = -2\alpha_2$$

or

$$\alpha_3 = -2\frac{\alpha_2}{3} = -4\frac{\alpha_1}{3}$$

There is no assurance that the value of any α is identically zero. In fact, the solution for the α's may be written as

$$\begin{bmatrix} \alpha_1 \\ \alpha_2 \\ \alpha_3 \end{bmatrix} = \beta \begin{bmatrix} 1 \\ 2 \\ -4/3 \end{bmatrix}$$

where β may be any scalar, not necessarily zero. The column vectors in the matrix A form a linearly dependent set, and any larger set containing these vectors as a subset forms a linearly dependent set.

5.3 LINEAR COMBINATIONS

If the vectors $A_1, A_2, A_3 \cdots A_n$ are linearly dependent, then at least one of them can be expressed as a linear combination of any or all of the others. This is easily observed by letting the scalar α_k be nonzero in a rearrangement of Eq. (5.1)

$$A_k = -\frac{\alpha_1}{\alpha_k}A_1 - \frac{\alpha_2}{\alpha_k}A_2 - \frac{\alpha_3}{\alpha_k}A_3 - \cdots - \frac{\alpha_n}{\alpha_k}A$$

which expresses A_k as a linear combination of all vectors that are associated with a nonzero value of α. Moreover, this indicates that if one vector in a set of vectors can be expressed as a linear combination of the others, the vectors are linearly dependent.

5.4 THE RANK OF A MATRIX

The rank of a matrix A, written as $r(A)$, is the largest value of r for which there exists an $r \times r$ submatrix of A that is nonsingular.

Example: The matrix

$$A = \begin{bmatrix} 1 & 2 & 3 \\ 4 & 5 & 6 \\ 7 & 8 & 9 \end{bmatrix}$$

is singular because det $A = 0$, as the reader may verify. This matrix cannot have a rank of 3, that is, $r(A) \neq 3$. However, it is easy to find a 2×2 submatrix, for example,

$$B = \begin{bmatrix} 1 & 2 \\ 4 & 5 \end{bmatrix}$$

where the determinant is not equal to zero (in this case, det $B = -3$). Thus the rank of the matrix A is 2 ($r(A) = 2$).

The matrix A may be $m \times n$, in which case the largest possible square matrix is $n \times n$. Any $n \times n$ matrix, whether formed from a matrix A which is $m \times n$ or not, has a rank $r = n$ as long as it possesses a nonvanishing determinant.

Example: The 3×4 matrix

$$A = \begin{bmatrix} 1 & -1 & 1 & 2 \\ 2 & -2 & 2 & 4 \\ 1 & 2 & 3 & 4 \end{bmatrix}$$

has a row that is a constant multiple of another row (row two is twice row one). This matrix possesses four submatrices:

$$\begin{bmatrix} 1 & -1 & 1 \\ 2 & -2 & 2 \\ 1 & 2 & 3 \end{bmatrix} \quad \begin{bmatrix} 1 & -1 & 2 \\ 2 & -2 & 4 \\ 1 & 2 & 4 \end{bmatrix}$$

$$\begin{bmatrix} 1 & 1 & 2 \\ 2 & 2 & 4 \\ 1 & 3 & 4 \end{bmatrix} \quad \begin{bmatrix} -1 & 1 & 2 \\ -2 & 2 & 4 \\ 3 & 3 & 4 \end{bmatrix}$$

Each of these submatrices is singular; the determinants of each of them has a value equal to zero because, in each of them, the second row is a constant multiple of the first row; that is, the second row is a linear combination of the first row. Thus the rank of the matrix A cannot be equal to 3; that is, $r(A) \neq 3$.

However, it is easy to find a 2×2 submatrix of A that is not singular. For example, take the elements at the lower left:

$$\det \begin{bmatrix} 2 & -2 \\ 1 & 2 \end{bmatrix} = 6$$

This indicates that $r(A) = 2$.

5.5 RELATIONSHIP BETWEEN RANK AND LINEAR DEPENDENCE

If one row of a matrix is a linear combination of any of the other rows, the value of its determinant is zero.† This fact is intimately related to the rank of the matrix. An nth-order matrix must have a rank less than n if the matrix is singular; looking at this in another way, an nth-order matrix with a rank that is less than n must be singular.

Now any set of row vectors that possesses a subset of linearly dependent row vectors must, itself, be a linearly dependent set. This means that if some of the row vectors are linearly dependent, all of them are linearly dependent.

However, by deleting the vectors that cause the linear combination and hence the linear dependency, it is possible to find a subset of linearly independent row vectors, and the determinant of such a subset will not be zero, because a nonvanishing determinant requires a linearly independent set of rows. Thus it is seen that the rank of a matrix is determined by the number of linear independent rows.

The same conclusions apply to the column vectors in a matrix. Suppose an $m \times n$ rectangular matrix possesses exactly q linearly independent rows. Its rank will be q and any square matrices of order $q + 1$ will be singular. But the determinant of a matrix is equal to the determinant of its transpose.‡ So in the transpose, there must be one submatrix of order q that is not singular, and all submatrices of order $q + 1$ are singular. Because the rows of a matrix are the columns of its transpose, it is observed that the the maximum number of linearly independent rows must be equal to the maximum number of independent columns.

5.6 CONDITIONS FOR THE SOLUTION OF SIMULTANEOUS EQUATIONS

A set of simultaneous, linear algebraic equations

$$AX = B \qquad (5.2)$$

is termed *consistent* if there is a solution and *inconsistent* if there is no solution. If B is not null, the set of equations is said to be *nonhomogeneous*, and if B is null; that is, if

$$AX = 0 \qquad (5.3)$$

the set of equations is said to be *homogeneous*.

---

†See rule five in Section 2.4.
‡See rule eight in Section 2.4.

The nth-order square matrix

$$
A = \begin{bmatrix}
a_{11} & a_{12} & a_{13} & \cdots & a_{1n} \\
a_{21} & a_{22} & a_{23} & \cdots & a_{2n} \\
a_{31} & a_{32} & a_{33} & \cdots & a_{3n} \\
& & \cdots & & \\
a_{n1} & a_{n2} & a_{n3} & \cdots & a_{nn}
\end{bmatrix}
\tag{5.4}
$$

may be augmented by adding the column vector B:

$$
A^a = \begin{bmatrix}
a_{11} & a_{12} & a_{13} & \cdots & a_{1n} & b_1 \\
a_{21} & a_{22} & a_{23} & \cdots & a_{2n} & b_2 \\
a_{31} & a_{32} & a_{33} & \cdots & a_{3n} & b_3 \\
& & \cdots & & & \\
a_{n1} & a_{n2} & a_{n3} & \cdots & a_{nn} & b_n
\end{bmatrix}
\tag{5.5}
$$

and it is seen that the order of A^a is $n \times (n + 1)$. Observe, however, that if the rank of A is n, then the rank of A^a must also be n. Also observe that it is clearly impossible for the rank of A to exceed the rank of A^a. Finally, note that the rank of A^a may never exceed the rank of A by more than one, because the vector B can add no more than one linearly independent column.

The summary chart in Fig. 5.1 has been constructed to account concisely for the types of equations, the conditions of rank, and the form of solutions. The chart pertains to both nonhomogeneous and homogeneous systems of equations, although further remarks concerning the solution of homogeneous equations are presented in Section 5.9. The summary chart is now used in conjunction with several examples pertaining to the nonhomogeneous cases.

5.7 EXAMPLES: NONHOMOGENEOUS CASES

Example: The system of equations

$$
\begin{aligned}
4x_1 - x_2 &= 12 \\
-x_1 + 5x_2 - 2x_3 &= 0 \\
- 2x_2 + 4x_3 &= -8
\end{aligned}
$$

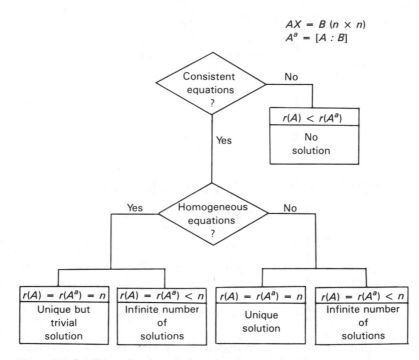

Figure 5.1 Conditions for the solution of n simultaneous, linear algebraic equations with coefficient matrix A and augmented matrix A^a showing types of equations, conditions of rank, and forms of solution.

has the coefficient matrix

$$A = \begin{bmatrix} 4 & -1 & 0 \\ -1 & 5 & -2 \\ 0 & -2 & 4 \end{bmatrix}$$

with $r(A) = 3$ because

$$\det A = 80 + 0 + 0 - 0 - 16 - 4 = 60$$

The system has the augmented matrix

$$A^a = \begin{bmatrix} 4 & -1 & 0 & 12 \\ -1 & 5 & -2 & 0 \\ 0 & -2 & 4 & -8 \end{bmatrix}$$

with $r(A^a) = 3$. Figure 5.1 indicates that there is a unique solution, because

$r(A) = r(A^a) = n = 3$. The solution is obtained via the inverse of A which the reader may verify:

$$X = \begin{bmatrix} x_1 \\ x_2 \\ x_3 \end{bmatrix} = \begin{bmatrix} 4/15 & 1/15 & 1/30 \\ 1/15 & 4/15 & 2/15 \\ 1/30 & 2/15 & 19/60 \end{bmatrix} \begin{bmatrix} 12 \\ 0 \\ -8 \end{bmatrix} = \begin{bmatrix} 44/15 \\ -4/15 \\ -32/15 \end{bmatrix}$$

Example: The system of equations

$$x_1 - x_2 + x_3 = 2$$
$$2x_1 - 2x_2 + 4x_3 = 8$$
$$x_1 - x_2 + 3x_3 = 6$$

has the coefficient matrix

$$A = \begin{bmatrix} 1 & -1 & 1 \\ 2 & -2 & 4 \\ 1 & -1 & 3 \end{bmatrix}$$

which is singular, and therefore, its rank is $r(A) = 2$ because a 2×2 submatrix can be found that is not singular. The system has the augmented matrix

$$A^a = \begin{bmatrix} 1 & -1 & 1 & 2 \\ 2 & -2 & 4 & 8 \\ 1 & -1 & 3 & 6 \end{bmatrix}$$

from which four 3×3 matrices may be formed by taking the combination of four columns, three at a time. One of these is A, which is singular. The others,

$$\begin{bmatrix} 1 & -1 & 2 \\ 2 & -2 & 8 \\ 1 & -1 & 6 \end{bmatrix} \quad \begin{bmatrix} -1 & 1 & 2 \\ -2 & 4 & 8 \\ -1 & 3 & 6 \end{bmatrix} \quad \begin{bmatrix} 1 & 1 & 2 \\ 2 & 4 & 8 \\ 1 & 3 & 6 \end{bmatrix}$$

are also singular, which means that $r(A^a) = 2$. Because $r(A) = r(A^a) = 2 < 3$, the system possesses an infinite number of solutions with $n - r = 3 - 2 = 1$ arbitrary constant.

The solutions may be obtained by first subtracting the first equation from the third:

$$2x_3 = 4$$
$$x_3 = 2$$

and with $x_3 = 2$ in all the equations, in each case one finds that $x_1 = x_2$, and, no matter what values are assigned to x_1, if $x_1 = x_2$ and $x_3 = 2$, the three values will satisfy the system of equations. If β designates the arbitrary parameter, then

$$X = \begin{bmatrix} x_1 \\ x_2 \\ x_3 \end{bmatrix} = \begin{bmatrix} \beta \\ \beta \\ 2 \end{bmatrix}$$

which holds for all values of β, including zero.

Example: The system of equations

$$4x_1 + 9x_2 + 3x_3 = 6$$

$$2x_1 + 3x_2 + x_3 = 2$$

$$2x_1 + 6x_2 + 2x_3 = 7$$

has the coefficient matrix

$$A = \begin{bmatrix} 4 & 9 & 3 \\ 2 & 3 & 1 \\ 2 & 6 & 2 \end{bmatrix}$$

that is singular and has a rank $r(A) = 2$, because a 2×2 submatrix can be found that is nonsingular. The system possesses the augmented matrix:

$$A^a = \begin{bmatrix} 4 & 9 & 3 & 6 \\ 2 & 3 & 1 & 2 \\ 2 & 6 & 2 & 7 \end{bmatrix}$$

that has a 3×3 submatrix such that

$$\det \begin{bmatrix} 4 & 9 & 6 \\ 2 & 3 & 2 \\ 2 & 6 & 7 \end{bmatrix} = -18$$

This indicates that $r(A^a) = 3$ and that $r(A) < r(A^a)$ or $2 < 3$ and that the system has no solution.

5.8 HOMOGENEOUS EQUATIONS

The system of equations

$$AX = 0 \tag{5.6}$$

is satisfied if either A or X or both are null. The trivial solution

$$
X = \begin{bmatrix} x_1 \\ x_2 \\ x_3 \\ \cdots \\ x_n \end{bmatrix} = \begin{bmatrix} 0 \\ 0 \\ 0 \\ \cdots \\ 0 \end{bmatrix}
$$

is presumed to be of no interest and is assured to evolve if A is nonsingular.

A nontrivial solution of Eq. (5.6) can occur if A is singular. This means that $r(A) < n$. Hence there are an infinite number of nontrivial solutions if the equations are consistent.

Example: The system of equations

$$4x_1 + 9x_2 + 3x_3 = 0$$

$$2x_1 + 3x_2 + x_3 = 0$$

$$2x_1 + 6x_2 + 2x_3 = 0$$

has the coefficient matrix

$$
A = \begin{bmatrix} 4 & 9 & 3 \\ 2 & 3 & 1 \\ 2 & 6 & 2 \end{bmatrix}
$$

that is singular and of rank $r(A) = 2$. The augmented matrix also has a rank of 2, $r(A^a) = 2$ because all three other 3×3 matrices formed with the column of zeroes are singular. Hence the system is consistent, because $r(A) = r(A^a) = 2 < n = 3$ and an infinite number of solutions exist.

Subtract the second equation from the third and get x_3 in terms of x_2:

$$3x_2 + x_3 = 0$$

$$x_3 = -3x_2$$

Then put this in the first equation and find that

$$4x_1 + 9x_2 + 3(-3x_2) = 0$$

or $x_1 = 0$.

If the value of x_2 is arbitrary, say $x_2 = \gamma$, then the solution vector is

$$X = \begin{bmatrix} x_1 \\ x_2 \\ x_3 \end{bmatrix} = \gamma \begin{bmatrix} 0 \\ 1 \\ 3 \end{bmatrix}$$

which holds for any and all values of γ.

Example: The system of equations

$$4x_1 - 2x_2 = 0$$
$$-2x_1 + 6x_2 = 0$$

has the coefficient matrix

$$A = \begin{bmatrix} 4 & -2 \\ -2 & 6 \end{bmatrix}$$

whose determinant is equal to 20. A Cramer's rule solution yields the trivial solution

$$x_1 = \frac{\begin{vmatrix} 0 & -2 \\ 0 & 6 \end{vmatrix}}{\begin{vmatrix} 4 & -2 \\ -2 & 6 \end{vmatrix}} = \frac{0}{20} = 0$$

and

$$x_2 = \frac{\begin{vmatrix} 4 & 0 \\ -2 & 0 \end{vmatrix}}{20} = \frac{0}{20} = 0$$

5.9 EQUIVALENCE OF MATRICES

Two matrices whose elements are real or complex numbers are said to be equivalent if and only if each can be transformed into the other by means of elementary transformations. From this, it is easy to see that two matrices can only be equivalent if they have the same rank and the same order.

5.10 GAUSSIAN ELIMINATION

Consider the set of three simultaneous, linear algebraic equations

$$2x_1 + x_2 + x_3 = 7$$

$$4x_1 + x_2 + x_3 = 9$$

$$2x_1 + 3x_2 + 2x_3 = 14$$

which is in the form of Eq. (5.2):

$$AX = B$$

with the coefficient matrix

$$A = \begin{bmatrix} 2 & 1 & 1 \\ 4 & 1 & 1 \\ 2 & 3 & 2 \end{bmatrix}$$

The solution of this nonhomogeneous consistent set exists and is unique. It may be established by a systematic procedure known as *Gaussian elimination*.

If the first equation is multiplied by 2 and then subtracted from the second, the result is

$$2x_1 + x_2 + x_3 = 7$$

$$- x_2 - x_3 = -5$$

$$2x_1 + 3x_2 + 2x_3 = 14$$

Here, the multiplier 2 is called the pivot, and it is observed that the x_1 term in the second equation has been eliminated.

The x_1 term may be eliminated from the third equation by selecting the number 1 as the pivot, multiplying the first equation by 1 (actually a matter of procedure), and subtracting the result from the third equation. This procedure yields

$$2x_1 + x_2 + x_3 = 7$$

$$-x_2 - x_3 = -5$$

$$2x_2 + x_3 = 7$$

and it is observed that the first term in both the second and third equations involves x_2.

The elimination process is concluded by selecting the number -2 as the pivot, multiplying the second equation by -2, and then subtracting the result from the third equation. This procedure provides

$$2x_1 + x_2 + x_3 = 7$$
$$-x_2 - x_3 = -5$$
$$-x_3 = -3$$

and at this point it is easy to solve for x_1, x_2, and x_3 by a procedure known as *back substitution*:

$$x_3 = 3$$

$$x_2 = 5 - x_3 = 5 - 3 = 2$$

and

$$2x_1 = 7 - x_2 - x_3 = 7 - 2 - 3 = 2$$

so that

$$x_1 = 1$$

An identical result could have been achieved by multiplying the system of equations by three elementary transformation matrices of the third kind, which is designated here with a double subscript to indicate the order of employment as premultipliers

$$E_{33}E_{32}E_{31}AX = E_{33}E_{32}E_{31}B$$

where

$$E_{31} = \begin{bmatrix} 1 & 0 & 0 \\ -2 & 1 & 0 \\ 0 & 0 & 1 \end{bmatrix}$$

$$E_{32} = \begin{bmatrix} 1 & 0 & 0 \\ 0 & 1 & 0 \\ -1 & 0 & 1 \end{bmatrix}$$

and

$$E_{33} = \begin{bmatrix} 1 & 0 & 0 \\ 0 & 1 & 0 \\ 0 & 2 & 1 \end{bmatrix}$$

This makes

$$E_{33}E_{32}E_{31} = \begin{bmatrix} 1 & 0 & 0 \\ 0 & 1 & 0 \\ 0 & 2 & 1 \end{bmatrix}\begin{bmatrix} 1 & 0 & 0 \\ 0 & 1 & 0 \\ -1 & 0 & 1 \end{bmatrix}\begin{bmatrix} 1 & 0 & 0 \\ -2 & 1 & 0 \\ 0 & 0 & 1 \end{bmatrix}$$

$$= \begin{bmatrix} 1 & 0 & 0 \\ 0 & 1 & 0 \\ -1 & 2 & 1 \end{bmatrix}\begin{bmatrix} 1 & 0 & 0 \\ -2 & 1 & 0 \\ 0 & 0 & 1 \end{bmatrix} = \begin{bmatrix} 1 & 0 & 0 \\ -2 & 1 & 0 \\ -5 & 2 & 1 \end{bmatrix}$$

and

$$E_{33}E_{32}E_{31}A = \begin{bmatrix} 1 & 0 & 0 \\ -2 & 1 & 0 \\ -5 & 2 & 1 \end{bmatrix}\begin{bmatrix} 2 & 1 & 1 \\ 4 & 1 & 1 \\ 2 & 3 & 2 \end{bmatrix} = \begin{bmatrix} 2 & 1 & 1 \\ 0 & -1 & -1 \\ 0 & 0 & -1 \end{bmatrix}$$

It is even easier to obtain

$$E_{33}E_{32}E_{31}B = \begin{bmatrix} 1 & 0 & 0 \\ -2 & 1 & 0 \\ -5 & 2 & 1 \end{bmatrix}\begin{bmatrix} 7 \\ 9 \\ 14 \end{bmatrix} = \begin{bmatrix} 7 \\ -5 \\ -3 \end{bmatrix}$$

and the solution strategy is thus confirmed; both A and B may be pre-multiplied by a sequence of elementary transformation matrices to yield a new system

$$UX = C \tag{5.7}$$

where $U = E_{33}E_{32}E_{31}A$ is an upper triangular matrix and $C = E_{33}E_{32}E_{31}B$ is a new column vector of constants. The system of Eq. (5.7) may be obtained by backward substitution, although nothing really prohibits a solution determined by a matrix inversion

$$X = U^{-1}C$$

Observe that, in general, through the use of n elementary transformations of the third kind,

$$E_{3n} \cdots E_{33}E_{32}E_{31}A = UX = E_{3n} \cdots E_{33}E_{32}E_{31}B = C$$

a new representation springs forth

$$TAX = UX = TB = C \tag{5.8}$$

where

$$T = E_{3n} \cdots E_{33}E_{32}E_{31}$$

If A is to be decomposed into a pair of triangular matrices, L which is lower triangular and U which is upper triangular, one may speak of a triangular decomposition of A or an LU factorization of A. If, as indicated by Eq. (5.8) and $A = LU$, it is easy to see that

$$TAX = TLUX = UX$$

$$T^{-1}TAX = T^{-1}TLUX = T^{-1}UX$$

and

$$L = T^{-1} = E_{31}^{-1}E_{32}^{-1}E_{33}^{-1} \cdots E_{3n}^{-1} \tag{5.9}$$

Example: In the example that has been used to illustrate the procedure to be followed in Gaussian elimination, suppose it is desired to find L. All that need be done is to take the inverses of the three elementary transformation matrices and string them together in reverse order. This is not that difficult when the matrices are of small order, and the procedure is assisted greatly by the fact that the determinant of a triangular matrix is equal to the product of the elements along the principal diagonal.

$$E_{31}^{-1} = \begin{bmatrix} 1 & 0 & 0 \\ 2 & 1 & 0 \\ 0 & 0 & 1 \end{bmatrix}$$

$$E_{32}^{-1} = \begin{bmatrix} 1 & 0 & 0 \\ 0 & 1 & 0 \\ 1 & 0 & 1 \end{bmatrix}$$

and

$$E_{33}^{-1} = \begin{bmatrix} 1 & 0 & 0 \\ 0 & 1 & 0 \\ 0 & -2 & 1 \end{bmatrix}$$

Thus

$$L = E_{31}^{-1}E_{32}^{-1}E_{33}^{-1} = \begin{bmatrix} 1 & 0 & 0 \\ 2 & 1 & 0 \\ 0 & 0 & 1 \end{bmatrix}\begin{bmatrix} 1 & 0 & 0 \\ 0 & 1 & 0 \\ 1 & 0 & 1 \end{bmatrix}\begin{bmatrix} 1 & 0 & 0 \\ 0 & 1 & 0 \\ 0 & -2 & 1 \end{bmatrix}$$

$$= \begin{bmatrix} 1 & 0 & 0 \\ 2 & 1 & 0 \\ 1 & 0 & 1 \end{bmatrix}\begin{bmatrix} 1 & 0 & 0 \\ 0 & 1 & 0 \\ 0 & -2 & 1 \end{bmatrix} = \begin{bmatrix} 1 & 0 & 0 \\ 2 & 1 & 0 \\ 1 & -2 & 1 \end{bmatrix}$$

and

$$LU = \begin{bmatrix} 1 & 0 & 0 \\ 2 & 1 & 0 \\ 1 & -2 & 1 \end{bmatrix} \begin{bmatrix} 2 & 1 & 1 \\ 0 & -1 & -1 \\ 0 & 0 & -1 \end{bmatrix} = \begin{bmatrix} 2 & 1 & 1 \\ 4 & 1 & 1 \\ 2 & 3 & 2 \end{bmatrix}$$

or $LU = A$ as required.

Observe that L is indeed lower triangular and that its off-diagonal elements are precisely the pivots used in the elementary transformation matrices used to generate U. Thus L may be formed without taking inverses of the elementary transformation matrices.

All the foregoing is pointing to an efficient computational method employing Gaussian elimination via an LU factorization. There are three ways of reducing the coefficient matrix A to an LU product. These are discussed in Section 5.12 after some material pertaining to triangular matrices is presented.

5.11 TRIANGULAR MATRICES

A square matrix is upper triangular if $a_{ij} = 0$ for all $i > j$. It is strictly upper triangular if its diagonal elements are zero and unit upper triangular if its diagonal elements are all equal to unity. A square matrix is lower triangular if $a_{ij} = 0$ for all $i < j$, strictly lower triangular if its diagonal elements are zero, and unit lower triangular if its diagonal elements are all equal to unity. The determinant of an upper or lower triangular matrix is equal to the product of the elements that compose the principal diagonal.

Example: The matrix

$$U = \begin{bmatrix} 1 & 1 & 2 & 3 \\ 0 & 2 & -1 & 1 \\ 0 & 0 & 2 & 4 \\ 0 & 0 & 0 & -1 \end{bmatrix}$$

is upper triangular (not strictly nor unit upper triangular) because all $u_{ij} = 0$ for $i > j$. Its determinant is det $U = -4$, and this can be confirmed by an expansion down the first column:

$$\det U = (1) \det \begin{bmatrix} 2 & -1 & 1 \\ 0 & 2 & 4 \\ 0 & 0 & -1 \end{bmatrix} = -4$$

Consider the nth-order upper triangular matrix U and suppose that it can be partitioned into

$$U = \begin{bmatrix} A & B \\ 0 & D \end{bmatrix}$$

where A is of order m and D is of order $n - m$. Equation (4.11) indicates that if A and D are nonsingular, U possesses an inverse with $P = A^{-1}B$ and $Q = D$:

$$U^{-1} = \left[\begin{array}{c|c} A^{-1} & -A^{-1}BD^{-1} \\ \hline 0 & D^{-1} \end{array} \right] \tag{5.10}$$

Here, U^{-1} is also upper triangular, and because U^{-1} cannot exist if U is singular, the determinant of U cannot be zero. Thus U cannot be strictly upper triangular.

Example: The upper triangular matrix

$$U = \begin{bmatrix} 1 & 2 & -1 & 3 & 4 \\ 0 & 2 & 1 & 1 & 1 \\ 0 & 0 & 3 & -1 & 0 \\ 0 & 0 & 0 & -1 & 4 \\ 0 & 0 & 0 & 0 & 4 \end{bmatrix}$$

does possess an inverse, because all elements on the principal diagonal are nonzero. Select A

$$A = \begin{bmatrix} 1 & 2 & -1 \\ 0 & 2 & 1 \\ 0 & 0 & 3 \end{bmatrix}$$

that is also upper triangular and has an inverse

$$A^{-1} = \begin{bmatrix} 1 & -1 & 2/3 \\ 0 & 1/2 & -1/6 \\ 0 & 0 & 1/3 \end{bmatrix}$$

which can be verified by the reader. With A so selected,

$$B = \begin{bmatrix} 3 & 4 \\ 1 & 1 \\ -1 & 0 \end{bmatrix}$$

and

$$Q = D = \begin{bmatrix} -1 & 4 \\ 0 & 4 \end{bmatrix}$$

so that

$$Q^{-1} = D^{-1} = \begin{bmatrix} -1 & 1 \\ 0 & 1/4 \end{bmatrix}$$

Then

$$A^{-1}BD^{-1} = \begin{bmatrix} 1 & -1 & 2/3 \\ 0 & 1/2 & -1/6 \\ 0 & 0 & 1/3 \end{bmatrix} \begin{bmatrix} 3 & 4 \\ 1 & 1 \\ -1 & 0 \end{bmatrix} \begin{bmatrix} -1 & 1 \\ 0 & 1/4 \end{bmatrix}$$

$$= \begin{bmatrix} 4/3 & 3 \\ 2/3 & 1/2 \\ -1/3 & 0 \end{bmatrix} \begin{bmatrix} -1 & 1 \\ 0 & 1/4 \end{bmatrix} = \begin{bmatrix} -4/3 & 25/12 \\ -2/3 & 19/24 \\ 1/3 & -1/3 \end{bmatrix}$$

and then, in accordance with Eq. (5.10),

$$U^{-1} = \begin{bmatrix} 1 & -1 & 2/3 & 4/3 & -25/12 \\ 0 & 1/2 & -1/6 & 2/3 & -19/24 \\ 0 & 0 & 1/3 & -1/3 & 1/3 \\ 0 & 0 & 0 & -1 & 1 \\ 0 & 0 & 0 & 0 & 1/4 \end{bmatrix}$$

that, like U, is upper triangular.

An nth-order lower triangular matrix L can be partitioned into

$$L = \begin{bmatrix} A & 0 \\ C & D \end{bmatrix}$$

If A is of order m, then D is of order $n - m$, and Eq. (4.11) indicates that, as long as A and D are nonsingular, L possesses an inverse. In this case, because B is null, $P = A^{-1}B = 0$, $Q = D$, $Q^{-1} = D^{-1}$, and L^{-1} is given by

$$L = \begin{bmatrix} A^{-1} & 0 \\ -D^{-1}CA^{-1} & D^{-1} \end{bmatrix} \qquad (5.11)$$

Example: Consider the lower triangular matrix

$$L = \begin{bmatrix} 2 & 0 & 0 & 0 \\ 1 & 3 & 0 & 0 \\ 2 & 3 & -1 & 0 \\ 4 & 2 & -1 & 2 \end{bmatrix}$$

Its inverse may be obtained by allowing A to be 2×2

$$A = \begin{bmatrix} 2 & 0 \\ 1 & 3 \end{bmatrix}$$

so that

$$A^{-1} = \begin{bmatrix} 1/2 & 0 \\ -1/6 & 1/3 \end{bmatrix}$$

$$C = \begin{bmatrix} 2 & 3 \\ 4 & 2 \end{bmatrix}$$

$$Q = D = \begin{bmatrix} -1 & 0 \\ -1 & 2 \end{bmatrix}$$

and

$$Q^{-1} = D^{-1} = \begin{bmatrix} -1 & 0 \\ -1/2 & 1/2 \end{bmatrix}$$

Then

$$D^{-1}CA^{-1} = \begin{bmatrix} -1 & 0 \\ -1/2 & 1/2 \end{bmatrix}\begin{bmatrix} 2 & 3 \\ 4 & 2 \end{bmatrix}\begin{bmatrix} 1/2 & 0 \\ -1/6 & 1/3 \end{bmatrix}$$

$$= \begin{bmatrix} -2 & -3 \\ 1 & -1/2 \end{bmatrix}\begin{bmatrix} 1/2 & 0 \\ -1/6 & 1/3 \end{bmatrix} = \begin{bmatrix} -1/2 & -1 \\ 7/12 & -1/6 \end{bmatrix}$$

and, in accordance with Eq. (5.11), the inverse of L is

$$L^{-1} = \begin{bmatrix} 1/2 & 0 & 0 & 0 \\ -1/6 & 1/3 & 0 & 0 \\ 1/2 & 1 & -1 & 0 \\ -7/12 & 1/6 & -1/2 & 1/2 \end{bmatrix}$$

which is also lower triangular. The reader may wish to verify that $LL^{-1} = I$.

5.12 TRIANGULAR DECOMPOSITIONS

The factorization

$$LU = A \tag{5.12}$$

is not unique, and to see this, insert any nonsingular diagonal matrix and its inverse between L and U:

$$LDD^{-1}U = LU = A = L'U'$$

where $L' = LD$ is lower triangular and $U' = D^{-1}U$ is upper triangular. Because $A = L'U' = LU$, the factorization $A = LU$ is not unique. Through the use of a nonsingular diagonal matrix, however, one may write

$$LDU = A \tag{5.13}$$

and the factorizations

$$L(DU) = A \tag{5.14a}$$

and

$$(LD)U = A \tag{5.14b}$$

are both unique.

The factorization given by Eq. (5.14a) associates the diagonal matrix with the upper triangular matrix and is called the *Doolittle decomposition* or the *Doolittle reduction*. It is obtained immediately from the Gausssian elimination process.

The factorization given by Eq. (5.14b) associates the diagonal matrix with the lower triangular matrix. It is called the *Crout decomposition* or *Crout reduction*. It has a computational advantage over the Doolittle reduction. Rounding or roundoff errors can be reduced through the use of double precision, and the Crout reduction allows this to be accomplished with less computer memory than would be required if the Doolittle reduction were employed.

If the matrix A is symmetrical and *positive definite*, the *Cholesky decomposition*

$$(LD^{1/2})(D^{1/2}L^{T}) = A \tag{5.15}$$

may be employed. This decomposition requires about half the number of multiplications required for both the Doolittle and the Crout reductions, and this is to be expected, because it is easy to obtain L^{T} from L.

When the reader is confronted with the system of equations

$$AX = B$$

he or she should resist the temptation to write automatically

$$X = A^{-1}B$$

and then proceed to calculate A^{-1}. The determination of A^{-1} requires about twice as many computations as either the Crout or the Doolittle reductions. This might, correctly, lead one to the inference that the inverse should never be computed unless it is required or needed.

5.13 THE CROUT REDUCTION

Let the matrix A that is $n \times n$ have elements a_{ij}. It is to be factored in accordance with Eq. (5.14b) which is the Crout decomposition. Both LD and U are $n \times n$, LD has elements l_{ij}, and U has elements u_{ij}.
The computational algorithm is for $k = 1, 2, 3, \cdots, n$

$$u_{kk} = 1 \tag{5.16a}$$

$$l_{ik} = a_{ik} - \sum_{m=1}^{k-1} l_{im} u_{mk} \qquad (i = k, k + 1, \cdots, n) \tag{5.16b}$$

$$u_{kj} = \frac{1}{l_{kk}} \left(a_{kj} - \sum_{m=1}^{k-1} l_{km} u_{mj} \right) \qquad (j = k + 1, k + 2, \cdots, n) \tag{5.16c}$$

The presence of the l_{kk} term in the denominator of Eq. (5.16c) poses no problem, because if the matrix A is nonsingular, l_{kk} can never be equal to zero.

Example: The Crout reduction is used to factor the matrix A that is 4×4:

$$A = \begin{bmatrix} 1 & 2 & 0 & -1 \\ 1 & 1 & -1 & 1 \\ 2 & -1 & 2 & 0 \\ 1 & 1 & 0 & -1 \end{bmatrix}$$

First observe from Eq. (5.16a) that

$$u_{11} = u_{22} = u_{33} = u_{44} = 1$$

Then use Eq. (5.16b) with $k = 1$. Notice that no summation need be taken, because $k - 1 = 0$:

$$l_{11} = a_{11} = 1$$
$$l_{21} = a_{21} = 1$$
$$l_{31} = a_{31} = 2$$

and

$$l_{41} = a_{41} = 1$$

Equation (5.16c) with $k = 1$ and no summation then gives

$$u_{12} = \frac{1}{l_{11}} a_{12} = 2$$

$$u_{13} = \frac{1}{l_{11}} a_{13} = 0$$

and

$$u_{14} = \frac{1}{l_{11}} a_{14} = -1$$

Next let $k = 2$. This time there is one term in the summations in Eqs. (5.16b) and (5.16c)

$$l_{22} = a_{22} - l_{21}u_{12} = 1 - 1(2) = -1$$
$$l_{32} = a_{32} - l_{31}u_{12} = -1 - 2(2) = -5$$
$$l_{42} = a_{42} - l_{41}u_{12} = 1 - 1(2) = -1$$

and then

$$u_{23} = \frac{1}{l_{22}} (a_{23} - l_{21}u_{13}) = -[-1 - 1(0)] = 1$$

and

$$u_{24} = \frac{1}{l_{22}}(a_{24} - l_{21}u_{14}) = -[1 - 1(-1)] = -2$$

For $k = 3$, there are two terms in the summations of Eqs. (5.16b) and (5.16c):

$$l_{33} = a_{33} - l_{31}u_{13} - l_{32}u_{23} = 2 - 2(0) - (-5)(1) = 7$$
$$l_{43} = a_{43} - l_{41}u_{13} - l_{42}u_{23} = 0 - 1(0) - (-1)(1) = 1$$

and

$$u_{34} = \frac{1}{l_{33}} (a_{34} - l_{31}u_{14} - l_{32}u_{24}) = \frac{1}{7}[0 - 2(-1) - (-5)(-2)] = -8/7$$

Finally, with $k = 4$, Eq. (5.16b) provides

$$l_{44} = a_{44} - l_{41}u_{14} - l_{42}u_{24} - l_{43}u_{34}$$
$$= -1 - (1)(-1) - (-1)(-2) - (1)(-8/7) = -6/7$$

All the foregoing may be summarized by taking the product $L'U$ where $L' = LD$:

$$L'U = \begin{bmatrix} 1 & 0 & 0 & 0 \\ 1 & -1 & 0 & 0 \\ 2 & -5 & 7 & 0 \\ 1 & -1 & 1 & -6/7 \end{bmatrix} \begin{bmatrix} 1 & 2 & 0 & -1 \\ 0 & 1 & 1 & -2 \\ 0 & 0 & 1 & -8/7 \\ 0 & 0 & 0 & 1 \end{bmatrix}$$

and the reader may verify that this product is indeed equal to the matrix A.

5.14 QUADRATIC FORMS AND POSITIVE DEFINITE MATRICES

Consider what is called the quadratic form of a symmetrical matrix A

$$Q(x) = X^T A X \tag{5.17}$$

where X is a column vector that is not null ($X \neq 0$). The matrix A is said to be positive definite if, for all $X \neq 0$,

$$X^T A X > 0 \tag{5.18a}$$

and positive semidefinite if

$$X^T A X \geq 0 \tag{5.18b}$$

A principle submatrix of an $n \times n$ matrix A:

$$A = \begin{bmatrix} a_{11} & a_{12} & a_{13} & a_{14} & \cdots & a_{1n} \\ a_{21} & a_{22} & a_{23} & a_{24} & \cdots & a_{2n} \\ a_{31} & a_{32} & a_{33} & a_{34} & \cdots & a_{3n} \\ & & \cdots & & & \\ a_{n1} & a_{n2} & a_{n3} & a_{n,4} & \cdots & a_{nn} \end{bmatrix}$$

is any submatrix of A whose principal diagonal is part of the principal diagonal of A. Thus a listing of the principal submatrices of A begins with a listing of the matrices containing the single elements

$$[a_{11}], [a_{22}], [a_{33}], \cdots, [a_{nn}]$$

going on through the 2×2 matrices, some of which are

$$\begin{bmatrix} a_{11} & a_{12} \\ a_{21} & a_{22} \end{bmatrix} \quad \begin{bmatrix} a_{11} & a_{13} \\ a_{31} & a_{33} \end{bmatrix} \quad \begin{bmatrix} a_{11} & a_{14} \\ a_{41} & a_{44} \end{bmatrix}$$

then picking up the 3×3's, one of which is

$$\begin{bmatrix} a_{11} & a_{13} & a_{15} \\ a_{31} & a_{33} & a_{35} \\ a_{51} & a_{53} & a_{55} \end{bmatrix}$$

and winding up with the matrix A itself.

It can be proved that a necessary and sufficient condition that the matrix A be positive definite is that the determinants of all the principal submatrices be positive.

In Section 3.11 it was shown that any square matrix A of order n could be made symmetric by a pre- or postmultiplication by its transpose. It can be shown that if A is symmetric

$$B = A^{\mathrm{T}}A$$

is positive semidefinite and if A is of rank n, $r(A) = n$, then it can be shown that B is positive definite. With $B = A^{\mathrm{T}}A$

$$Q(x) = X^{\mathrm{T}}BX = X^{\mathrm{T}}A^{\mathrm{T}}AX$$

If $Y = AX$, then

$$Q(x) = Y^{\mathrm{T}}Y = \sum_{k=1}^{n} y_{kk}^2 \geq 0$$

and this proves that B is at least positive semidefinite,

But if $r(A) = n$, then the requirement that $Y = AX \neq 0$ implies that $X \neq 0$. Hence

$$Q(x) = X^{\mathrm{T}}BX > 0$$

and B is positive definite.

If A that is square, symmetrical, nonsingular, and of order $n \times n$ is positive definite, there is a unique lower triangular matrix that permits a factorization $A = LL^{\mathrm{T}}$. Because both L and L^{T} contain the square root of a diagonal matrix, the requirement that A be positive definite is crucial. Otherwise, square roots of negative numbers could come forth as part of the factorization procedure, and this would violate the premise that L and L^{T} contain only real elements.

5.15 THE CHOLESKY REDUCTION

Let the matrix A that is $n \times n$ and symmetric have elements a_{ij}. It is to be factored in accordance with Eq. (5.15) which is the Cholesky decomposition. Here L is also $n \times n$ and is lower triangular with elements l_{ij}.

The computational algorithm is for $k = 1, 2, 3, \ldots, n$

$$l_{ki} = \frac{1}{l_{ii}} \left(a_{ki} - \sum_{m=1}^{i-1} l_{im} l_{km} \right) \quad (i = 1, 2, \cdots, k - 1) \quad (5.19a)$$

$$l_{kk} = \sqrt{a_{kk} - \sum_{m=1}^{k-1} l_{km}^2} \quad (5.19b)$$

Notice that when $k = 1$, Eqs. (5.19a) and (5.19b) provide the same result, the element l_{11}.

Example: The Cholesky decomposition for the matrix

$$A = \begin{bmatrix} 4 & -1 & 0 & -1 \\ -1 & 6 & -1 & -2 \\ 0 & -1 & 4 & -2 \\ -1 & -2 & -2 & 8 \end{bmatrix}$$

is to be determined. First, however, it is necessary to determine whether A, which is symmetric, is positive definite. This is done by evaluating the determinants of the four principal submatrices of A:

$$\det [4] = 4$$

$$\det \begin{bmatrix} 4 & -1 \\ -1 & 6 \end{bmatrix} = 23$$

$$\det \begin{bmatrix} 4 & -1 & 0 \\ -1 & 6 & -1 \\ 0 & -1 & 4 \end{bmatrix} = 88$$

and for the matrix A itself, a pivotal condensation (see Section 2.8) can be employed:

$$\det A = \frac{1}{16} \det \left[4 \begin{bmatrix} 6 & -1 & -2 \\ -1 & 4 & -2 \\ -2 & -2 & 8 \end{bmatrix} - \begin{bmatrix} -1 \\ 0 \\ -1 \end{bmatrix} [-1 \quad 0 \quad -1] \right]$$

$$\det A = \frac{1}{16} \det \left[\begin{bmatrix} 24 & -4 & -8 \\ -4 & 16 & -8 \\ -8 & -8 & 32 \end{bmatrix} - \begin{bmatrix} 1 & 0 & 1 \\ 0 & 0 & 0 \\ 1 & 0 & 1 \end{bmatrix} \right]$$

or

$$\det A = \frac{1}{16} \det \begin{bmatrix} 23 & -4 & -9 \\ -4 & 16 & -8 \\ -9 & -8 & 31 \end{bmatrix} = \frac{7568}{16} = 473$$

This shows that A is positive definite, and the Cholesky reduction may now proceed.

For $k = 1$, the only element required is l_{11}. This element may be evaluated from either of Eqs. (5.19). Use of Eq. (5.19b) gives

$$l_{11} = \sqrt{a_{11}} = 2.0000$$

for $k = 2$ and $i = 1$, Eq. (5.19a) yields

$$l_{21} = \frac{1}{l_{11}} a_{21} = 1/2(-1) = -0.5000$$

and from Eq. (5.19b)

$$l_{22} = \sqrt{a_{22} - (l_{21}^2)} = \sqrt{6 - (1/2)^2} = \sqrt{5.75} = 2.3979$$

For $k = 3$ and $i = 1$ and 2, Eq. (5.19a) provides

$$l_{31} = \frac{1}{l_{11}} a_{31} = 1/2(0) = 0$$

and

$$l_{32} = \frac{1}{l_{22}}(a_{32} - l_{21}l_{31}) = \frac{1}{2.3979}[-1 - (-0.5000)(0)] = -0.4170$$

Then from Eq. (5.19b),

$$l_{33} = \sqrt{a_{33} - (l_{31}^2) - (l_{32}^2)} = \sqrt{4 - (0)^2 - (0.4170)^2} = 1.9560$$

With $k = 4$ and $i = 1, 2,$ and 3, Eq. (5.19a) is used:

$$l_{41} = \frac{1}{l_{11}} a_{41} = 1/2(-1) = -0.5000$$

$$l_{42} = \frac{1}{l_{22}}(a_{42} - l_{21}l_{41}) = \frac{1}{2.3979}[-2 - (-0.5000)(-0.5000)] = -0.9383$$

and

$$l_{43} = \frac{1}{l_{33}}(a_{43} - l_{31}l_{41} - l_{32}l_{42})$$

or

$$l_{43} = \frac{1}{1.9560}[-2 - 0(-0.5000) - (-0.4170)(-0.9383)] = -1.2225$$

Finally, use of Eq. (5.19b) gives

$$l_{44} = \sqrt{a_{44} - (l_{41}^2) - (l_{42}^2) - (l_{43}^2)}$$

or

$$l_{44} = \sqrt{8 - (-0.5000)^2 - (-0.9383)^2 - (-1.2225)^2} = 2.3184$$

The matrix L is

$$L = \begin{bmatrix} 2.0000 & 0.0000 & 0.0000 & 0.0000 \\ -0.5000 & 2.3979 & 0.0000 & 0.0000 \\ 0.0000 & -0.4170 & 1.9560 & 0.0000 \\ -0.5000 & -0.9383 & -1.2225 & 2.3184 \end{bmatrix}$$

and its transpose is easy to see. The reader may wish to verify that the product $LL^T = A$.

5.16 THE FORWARD AND BACK SUBSTITUTION ALGORITHMS

The solution vector X in the system

$$AX = B$$

can be found by first factoring A

$$LUX = B$$

and then letting $Y = UX$ so that the system is reduced to

$$LY = B$$

The vector Y can be obtained via the inverse of L or through the use of a forward substitution procedure involving L and B. A similar procedure involving the inverse of U or a back substitution procedure then yields the solution vector X.

The termination of the LU factorization of a system of equations where A is $n \times n$ provides $LY = B$ where $Y = UX$ and

$$l_{11}y_1 = b_1$$

$$l_{21}y_1 + l_{22}y_2 = b_2$$

$$l_{31}y_1 + l_{32}y_2 + l_{33}y_3 = b_3$$

. . .

$$l_{n1}y_1 + l_{n2}y_2 + l_{n3}y_3 + \cdots l_{nn}y_n = b_4$$

The forward substitution computational algorithm for the determination of Y is easy to see. For $i = 1, 2, 3, \ldots, n$

$$y_i = \frac{1}{l_{ii}}\left(b_i - \sum_{m=1}^{i-1} l_{im}y_m\right) \tag{5.20}$$

When each y has been evaluated, the system $UX = Y$ is

$$u_{11}x_1 + u_{12}x_2 + u_{13}x_3 + \cdots u_{1n}x_n = y_1$$

$$u_{22}x_2 + u_{23}x_3 + \cdots u_{2n}x_n = y_2$$

$$u_{33}x_3 + \cdots u_{3n}x_n = y_3$$

. . .

$$u_{nn}x_n = y_n$$

and this time the computational algorithm represents a back substitution. For $k = n, n - 1, \ldots, 2, 1,$

$$x_k = \frac{1}{u_{kk}}\left(y_k - \sum_{m=k+1}^{n} u_{km}x_m\right) \tag{5.21}$$

Note that if this algorithm is being used in conjunction with the Crout reduction, the procedure is facilitated by the fact that all $u_{kk} = 1$.

Example: Consider the system of equations

$$x_1 + 2x_2 - x_4 = 1$$

$$x_1 + x_2 - x_3 + x_4 = 4$$

$$2x_1 - x_2 + 2x_3 = 6$$

$$x_1 + x_2 - x_4 = -1$$

In matrix form the system is represented by $AX = B$

$$\begin{bmatrix} 1 & 2 & 0 & -1 \\ 1 & 1 & -1 & 1 \\ 2 & -1 & 2 & 0 \\ 1 & 1 & 0 & -1 \end{bmatrix}\begin{bmatrix} x_1 \\ x_2 \\ x_3 \\ x_4 \end{bmatrix} = \begin{bmatrix} 1 \\ 4 \\ 6 \\ -1 \end{bmatrix}$$

The matrix A possesses a unique factorization, and in the example in Section 5.13 it is shown that the Crout reduction yields the LU factorization:

$$\begin{bmatrix} 1 & 0 & 0 & 0 \\ 1 & -1 & 0 & 0 \\ 2 & -5 & 7 & 0 \\ 1 & -1 & 1 & -6/7 \end{bmatrix} \begin{bmatrix} 1 & 2 & 0 & -1 \\ 0 & 1 & 1 & -2 \\ 0 & 0 & 1 & -8/7 \\ 0 & 0 & 0 & 1 \end{bmatrix}$$

The algorithm of Eq. (5.20) may be used to find the column vector Y in $LY = B$. With

$$B = \begin{bmatrix} 1 \\ 4 \\ 6 \\ -1 \end{bmatrix}$$

it is seen from Eq. (5.20) that

$$y_1 = 1$$

$$y_2 = \frac{1}{-1}[4 - 1(1)] = -3$$

$$y_3 = \frac{1}{7}[6 - 2(1) - (-5)(-3)] = -11/7$$

and

$$y_4 = \frac{1}{-6/7}[-1 - 1(1) - (-1)(-3) - (1)(-11/7)] = 4$$

Now with $UX = Y$, the algorithm of Eq. (5.21) can be used to find the column vector X:

$$x_4 = 4$$

$$x_3 = -11/7 - (-8/7)(4) = 3$$

$$x_2 = -3 - 1(3) - (-2)(4) = 2$$

and

$$x_1 = 1 - 2(2) - 0(3) - (-1)(4) = 1$$

Thus

$$X = \begin{bmatrix} 1 \\ 2 \\ 3 \\ 4 \end{bmatrix}$$

and the reader may wish to verify that this is correct.

5.17 EXERCISES

5.1 Determine whether the vectors A, B, and C are linearly independent:

$$A = \begin{bmatrix} 1 \\ 2 \\ -1 \end{bmatrix} \quad B = \begin{bmatrix} 2 \\ 4 \\ -2 \end{bmatrix} \quad C = \begin{bmatrix} 1 \\ -2 \\ 3 \end{bmatrix}$$

5.2 Are the vectors E, F, G, and H linearly independent?

$$E = \begin{bmatrix} 1 \\ 2 \\ 3 \\ 4 \end{bmatrix} \quad F = \begin{bmatrix} 1 \\ 0 \\ 1 \\ 3 \end{bmatrix} \quad G = \begin{bmatrix} 1 \\ 2 \\ -1 \\ 0 \end{bmatrix} \quad H = \begin{bmatrix} 4 \\ 3 \\ 2 \\ 1 \end{bmatrix}$$

5.3 Determine the rank of

$$A = \begin{bmatrix} 1 & 2 & -1 & 2 & 4 \\ 2 & 3 & -1 & 0 & 1 \\ -1 & -2 & 3 & -2 & -4 \\ 0 & 1 & 2 & -1 & 0 \end{bmatrix}$$

5.4 Determine the rank of

$$A = \begin{bmatrix} 4 & -1 & 3 & 0 & 2 \\ 1 & 2 & -1 & 1 & 1 \\ 3 & 1 & -1 & 0 & 1 \\ -1 & 0 & 2 & -2 & 2 \\ 1 & 2 & 3 & -1 & 0 \end{bmatrix}$$

In exercises 5.5 through 5.9 the goal is to determine the nature of the solutions of the indicated set of simultaneous equations. In all of these the equations are to be solved, if possible, and the solutions are to be displayed as a column vector X.

5.5

$$x_1 + x_2 - 2x_3 = 4$$
$$2x_1 - x_2 + x_3 = 8$$
$$-2x_1 - 2x_2 + 4x_3 = -8$$

5.6

$$x_1 + 2x_2 + 3x_3 = 14$$
$$-x_1 - 2x_2 - 3x_3 = -15$$
$$x_1 + x_3 = 4$$

5.7

$$4x_1 - x_2 = 6$$
$$-x_1 + 6x_2 - 2x_3 = 4$$
$$- 2x_2 + 3x_3 = -2$$

5.8

$$2x_1 - x_2 + 4x_3 = 0$$
$$x_1 + x_2 + 2x_3 = 0$$
$$2x_1 + 2x_2 + 4x_3 = 0$$

5.9

$$4x_1 - 2x_2 - x_3 = 0$$
$$-2x_1 + 8x_2 - 3x_3 = 0$$
$$-x_1 - 3x_2 + 6x_3 = 0$$

Exercises 5.10 to 5.14 consider the system of equations

$$3x_1 - x_2 - x_3 = -4$$
$$-x_1 + 4x_2 - x_4 = 6$$
$$-x_1 + 2x_3 - x_4 = 2$$
$$-x_2 - x_3 + 4x_4 = 22$$

The system is to be solved by various methods:

5.10 Cramer's rule

5.11 Gauss elimination

5.12 Matrix inversion

5.13 The use of the Crout reduction

5.14 The use of the Cholesky reduction

CHAPTER SIX

ORTHOGONALITY AND COORDINATE TRANSFORMATIONS

6.1 INTRODUCTION

The concept of orthogonality has been briefly introduced in Section 1.7, where it was stated that two vectors are orthogonal (possess the property of orthogonality) if their scalar or dot product is equal to zero. In this chapter, this concept, which can be extended to matrices, is explored to a greater extent. The chapter also considers the transformation of coordinates, which is often requisite to the conduct of a successful engineering analysis.

6.2 ORTHOGONAL VECTORS

A vector in space may be described by its components that are projections in the three coordinate directions. In the Cartesian coordinate framework in Fig. 6.1a, where unit vectors in the positive x, y, and z coordinate directions are designated respectively by $\mathbf{i}$, $\mathbf{j}$, and $\mathbf{k}$, the vector P can be represented by

$$P = p_1\mathbf{i} + p_2\mathbf{j} + p_3\mathbf{k}$$

In a Cartesian coordinate system the coordinate axes are mutually perpendicular and are said to be orthogonal. Any line segment from the origin to a point $P(x, y, z)$ can be represented by a vector. For example, in Fig. 6.1b the three orthogonal axes are represented by x_1, x_2, and x_3. The line segment from the origin to the point P, OP can be represented in terms of its components and the unit vectors associated with the coor-

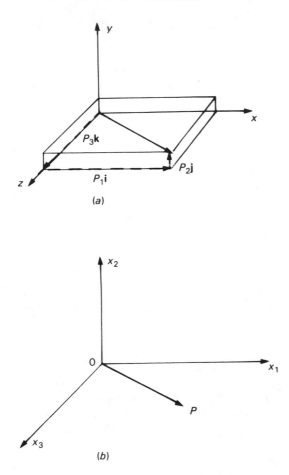

Figure 6.1 (*a*) A vector in a Cartesian coordinate system with coordinate axes *x*, *y*, and *z*; (*b*) the vector in a more general coordinate system with coordinate axes x_1, x_2, and x_3.

dinate axes. The vector *OP*, in this case, has a length or magnitude that is calculated using the Pythagorean theorem:

$$OP = \sqrt{x_1^2 + x_2^2 + x_3^2}$$

The directed line segment from the origin of the coordinate system to the point $P(x_1, x_2, x_3)$ is called a *vector*, and the vector is represented here, not in terms of coordinate and unit vectors, but as a 3 × 1 column of elements

$$X = \begin{bmatrix} x_1 \\ x_2 \\ x_3 \end{bmatrix}$$

or as a 1 × 3 row of elements

$$X^{\mathrm{T}} = \begin{bmatrix} x_1 & x_2 & x_3 \end{bmatrix}$$

Moreover, as a consequence of the definition of the scalar or dot product (see Section 1.7), the magnitude or length of the vector X may be defined as

$$|X| = (X^TX)^{1/2} = \sqrt{x_1^2 + x_2^2 + x_3^2}$$

The foregoing reasoning can be extended to an $n \times 1$ vector whose length or magnitude is

$$|X| = (X^TX)^{1/2} = \sqrt{x_1^2 + x_2^2 + x_3^2 + \cdots + x_n^2} \qquad (6.1)$$

and this is often referred to as the *norm of the vector*. If Y is also $n \times 1$, then the distance between X and Y is

$$d(X,Y) = d(Y,X) = \sqrt{\sum_{i=1}^{n}(x_i - y_i)^2} \qquad (6.2)$$

and it can be noted that if Y is null, it has no magnitude. In this event, Y is at the origin, and Eq. (6.2) reduces to Eq. (6.1).

A *normalized vector* is a vector having unit magnitude. It is obtained from another vector by dividing each element by its magnitude. Normalized vectors are often referred to as *unit vectors*, but this nomenclature is not used here.

Example: The vector

$$X = \begin{bmatrix} 1 \\ 2 \\ 3 \\ 4 \end{bmatrix}$$

has a magnitude

$$X = \sqrt{(1)^2 + (2)^2 + (3)^2 + (4)^2} = \sqrt{30}$$

and it can be normalized to

$$Y = \begin{bmatrix} 1/\sqrt{30} \\ 2/\sqrt{30} \\ 3/\sqrt{30} \\ 4/\sqrt{30} \end{bmatrix}$$

The vector Y has a magnitude

$$Y = \sqrt{(1/\sqrt{30})^2 + (2/\sqrt{30})^2 + (3/\sqrt{30})^2 + (4/\sqrt{30})^2}$$

$$= \sqrt{1/30 + 4/30 + 9/30 + 16/30} = \sqrt{30/30} = 1$$

6.3 THE SCALAR OR DOT PRODUCT REVISITED

The scalar or dot product has been defined in Section 1.7. Here, additional properties of such products are investigated. If both U and V are $n \times 1$ column vectors, the scalar or dot product $U{\cdot}V$ is defined as a scalar

$$U \cdot V = U^{\mathrm{T}}V = \sum_{i=1}^{n} u_i v_i \tag{6.3}$$

The scalar product is commutative

$$U \cdot V = V \cdot U = U^{\mathrm{T}}V = V^{\mathrm{T}}U$$

and distributive

$$(\alpha U + \beta V) \cdot W = \alpha(U \cdot W) + \beta(V \cdot W)$$

Moreover, as clearly shown by Eq. (6.3)

$$V \cdot V \geqslant 0$$

and if $V \cdot V = 0$, then V is null ($V = 0$).

The length of a vector as defined by Eq. (6.1) is

$$|X| = (X^{\mathrm{T}}X)^{1/2} = \sqrt{x_1^2 + x_2^2 + x_3^2 + \cdots + x_n^2}$$

and from this it is easy to see that $|X| \geqslant 0$ for all x, $|X| = 0$ if and only if all $x = 0$, and finally for all scalars α,

$$|\alpha X| = |\alpha||X|$$

6.4 THE CAUCHY–SCHWARTZ AND TRIANGLE INEQUALITIES

For the two $n \times 1$ column vectors column vectors U and V, it may be assumed that V is not null ($V \neq 0$), because if $V = 0$,

$$U \cdot V = U \cdot V = 0$$

The Cauchy–Schwartz inequality states that

$$|U\|V| \geqslant |U \cdot V| \tag{6.4}$$

and this can be proven by lettting the scalar products involving U and V be represented by

$$\alpha = V \cdot V$$

$$\beta = U \cdot V$$

and

$$\gamma = U \cdot U$$

If the inequality is to be valid, $\alpha\gamma$ must be greater than or equal to β^2 or $\alpha\gamma \geqslant \beta^2$. There is a real scalar such that

$$0 \leqslant (\delta V - U) \cdot (\delta V - U) = \alpha\delta^2 - 2\beta\delta + \gamma$$

In particular, if $\delta = \beta/\alpha$, then

$$0 \leqslant \alpha\left(\frac{\beta}{\alpha}\right)^2 - 2\beta\left(\frac{\beta}{\alpha}\right) + \gamma = \frac{\alpha\gamma - \beta^2}{\alpha}$$

and $\alpha\gamma - \beta^2 \geqslant 0$. Moreover, this shows that Eq. (6.4) holds.

The triangle inequality also concerns two vectors, X and Y:

$$|X + Y| \leqslant |X| + |Y| \tag{6.5}$$

Here $|X + Y|$ may be squared:

$$|X + Y|^2 = (X + Y) \cdot (X + Y) = X \cdot X + 2 X \cdot Y + Y \cdot Y$$

and this is less than or equal to

$$(|X| + |Y|)^2 = |X|^2 + 2|X||Y| + |Y|^2$$

and this indicates the validity of Eq. (6.5).

6.5 THE ANGLE AND THE PERPENDICULAR DISTANCE BETWEEN TWO VECTORS

Two vectors U and V are shown with a pair of coordinate axes in Fig. 6.2. The vector W can be obtained from a simple subtraction, $W = V - U$. The triangle indicated has sides U, V, and $W = V - U$, and in accordance with the law of cosines,

$$|W|^2 = |V - U|^2 = |U|^2 + |V|^2 - 2|U||V|\cos\theta$$

where θ is seen to be the angle between U and V.

The lengths or magnitudes may be expressed in terms of the scalar products

$$(V - U) \cdot (V - U) = V \cdot V - 2 U \cdot V + U \cdot U$$

$$= U \cdot U + V \cdot V - 2|U||V|\cos\theta$$

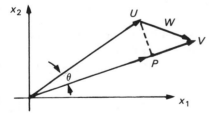

Figure 6.2 Four vectors in an x_1, x_2 coordinate system.

and a little algebra then shows that

$$2\, U \cdot V = 2|U||V|\cos\theta$$

Thus

$$\cos\theta = \frac{U \cdot V}{|U||V|} = \frac{U^{\mathrm{T}}V}{|U||V|} \tag{6.6a}$$

or

$$\theta = \text{arc}\cos\frac{U^{\mathrm{T}}V}{|U||V|} \tag{6.6b}$$

At this point the relationship between perpendicular vectors and the concept of orthogonality becomes quite apparent. If the angle $\theta = 90°$, then $\cos\theta = 0$, and $U \cdot V = 0$. Thus when vectors are perpendicular to each other, the vectors are orthogonal, and the dot or scalar product is equal to zero.

Figure 6.2 also shows a dashed line that is perpendicular to V and intersects V at point P such that the vector P is a scalar multiple of V:

$$P = \alpha V$$

The magnitude of the scalar is easily determined by noting that the vector $U - P = U - \alpha V$ is perpendicular to V. Thus

$$V \cdot (U - \alpha V) = 0$$

and a little algebra provides

$$V \cdot U - \alpha V \cdot V = 0$$

or

$$\alpha = \frac{V \cdot U}{V \cdot V}$$

and

$$P = \frac{V \cdot U}{V \cdot V}\,V \tag{6.7}$$

6.6 DIRECTION COSINES

An alternate description of a vector in space can be obtained by providing its magnitude and direction. The magnitude is described by Eq. (6.1), and for the case of the vector described in terms of its components and the unit vectors in the Cartesian framework shown in Fig. 6.3,

$$X = x_1\mathbf{x}_1 + x_2\mathbf{x}_2 + x_3\mathbf{x}_3$$

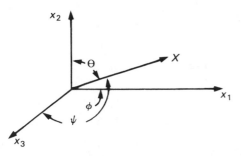

Figure 6.3 A vector in an x_1, x_2 and x_3 Cartesian coordinate system. The angles between the vector and the coordinate axes are called the direction angles.

The direction can be specified by the so-called direction angles ϕ, θ, and ψ which are measured, respectively, from the positive x_1, x_2, and x_3 directions to the vector.

It is almost always more convenient to represent the three angles in terms of the direction cosines, because the direction cosines can be prescribed in terms of the components and the magnitude of the vector by the following simple relationships:

$$\cos \phi = \frac{x_1}{|X|} \tag{6.8a}$$

$$\cos \theta = \frac{x_2}{|X|} \tag{6.8b}$$

and

$$\cos \psi = \frac{x_3}{|X|} \tag{6.8c}$$

It is, of course, easy to verify that the direction cosines are related by

$$\cos^2\phi + \cos^2\theta + \cos^2\psi = 1$$

which means that a knowledge of two direction cosines provides the value of the third except for the sign of the third.

The concept of the direction cosines may be extended to the $n \times 1$ vector X. In this case, according to Eqs. (6.8), there will be n direction cosines:

$$c_i = \sum_{i=1}^{n} \frac{x_i}{|X|}$$

6.7 THE LINEAR INDEPENDENCE OF A SET OF VECTORS

A very simple check for the linear dependence or independence of a set of n given vectors exists. Consider the pair of vectors A_1 and A_2 and attempt to form a linear combination of them:

$$B = \alpha_1 A_1 + \alpha_2 A_2$$

where the constants α_1 and α_2 can be uniquely determined and are not equal to zero.

To obtain α_1 and α_2, two scalar or dot products are taken:

$$B \cdot A_1 = \alpha_1 A_1 \cdot A_1 + \alpha_2 A_2 \cdot A_1$$

$$B \cdot A_2 = \alpha_1 A_1 \cdot A_2 + \alpha_2 A_2 \cdot A_2$$

and these may be written in matrix form as

$$\begin{bmatrix} A_1 \cdot A_1 & A_2 \cdot A_1 \\ A_1 \cdot A_2 & A_2 \cdot A_2 \end{bmatrix} \begin{bmatrix} \alpha_1 \\ \alpha_2 \end{bmatrix} = \begin{bmatrix} B \cdot A_1 \\ B \cdot A_2 \end{bmatrix}$$

If there is to be a unique solution for α_1 and α_2, it is necessary that the matrix

$$G = \begin{bmatrix} A_1 \cdot A_1 & A_2 \cdot A_1 \\ A_1 \cdot A_2 & A_2 \cdot A_2 \end{bmatrix}$$

be nonsingular; that is, det $G \neq 0$. The determinant is known as the *Gram determinant* or *Grammian*, which for n vectors, $A_1, A_2, A_3, \ldots, A_n$ can be written as

$$G = \begin{bmatrix} A_1 \cdot A_1 & A_2 \cdot A_1 & A_3 \cdot A_1 & \cdots & A_n \cdot A_1 \\ A_1 \cdot A_2 & A_2 \cdot A_2 & A_3 \cdot A_2 & \cdots & A_n \cdot A_2 \\ A_1 \cdot A_3 & A_2 \cdot A_3 & A_3 \cdot A_3 & \cdots & A_n \cdot A_3 \\ & & \cdots & & \\ A_1 \cdot A_n & A_2 \cdot A_n & A_3 \cdot A_n & \cdots & A_n \cdot A_n \end{bmatrix} \qquad (6.9a)$$

or alternatively as

$$G = \begin{bmatrix} A_1^T A_1 & A_2^T A_1 & A_3^T A_1 & \cdots & A_n^T A_1 \\ A_1^T A_2 & A_2^T A_2 & A_3^T A_2 & \cdots & A_n^T A_2 \\ A_1^T A_3 & A_2^T A_3 & A_3^T A_3 & \cdots & A_n^T A_3 \\ & & \cdots & & \\ A_1^T A_n & A_2^T A_n & A_3^T A_n & \cdots & A_n^T A_n \end{bmatrix} \qquad (6.9b)$$

It is now shown that if none of the vectors is parallel (colinear), which means that none is proportional to any of the others, det $G \neq 0$. This is done for the 2 × 2 case, and an easy extension to the $n \times n$ case can be established. Here

$$\det G = \det \begin{bmatrix} A_1 \cdot A_1 & A_2 \cdot A_1 \\ A_1 \cdot A_2 & A_2 \cdot A_2 \end{bmatrix} = |A_1|^2 |A_2|^2 - |A_1 \cdot A_2|^2$$

Because

$$A_1 \cdot A_2 = |A_1||A_2|\cos \theta$$

where θ is the angle between A_1 and A_2,

$$\begin{aligned} \det G &= |A_1|^2 |A_2|^2 - (|A_1||A_2| \cos \theta)^2 \\ &= |A_1|^2 |A_2|^2 (1 - \cos^2 \theta) \\ &= |A_1|^2 |A_2|^2 \sin \theta \end{aligned}$$

which is just twice the area of the triangle formed with A_1 and A_2 as adjacent sides with an included angle θ. Thus, as long as $\theta \neq 0$ (the vectors A_1 and A_2 are not colinear), $\det G \neq 0$, and the vectors A_1 and A_2 are linearly independent.

Thus a necessary and sufficient condition for the linear dependence of two vectors is the vanishing of the Grammian. For the linear dependence of n vectors, the Grammian, as given by Eqs. (6.9a) and (6.9b), must vanish.

6.8 ORTHOGONAL MATRICES

A square matrix A whose transpose is equal to its inverse

$$A^T = A^{-1}$$

is called an *orthogonal matrix*.

Example: The matrix

$$A = \begin{bmatrix} \cos \theta & -\sin \theta \\ \sin \theta & \cos \theta \end{bmatrix}$$

has the determinant

$$\det A = \cos^2 \theta + \sin^2 \theta = 1$$

the inverse

$$A^{-1} = \begin{bmatrix} \cos \theta & \sin \theta \\ -\sin \theta & \cos \theta \end{bmatrix}$$

and the transpose

$$A^T = \begin{bmatrix} \cos \theta & \sin \theta \\ -\sin \theta & \cos \theta \end{bmatrix}$$

The matrix A is orthogonal because $A^T = A^{-1}$.

Although two or more vectors may form an orthogonal set, they cannot form an orthogonal matrix unless they are orthonormal.

Example: The matrix A

$$A = \begin{bmatrix} 5 & -12 \\ 12 & 5 \end{bmatrix}$$

contains two column vectors

$$A_1 = \begin{bmatrix} 5 \\ 12 \end{bmatrix} \qquad A_2 = \begin{bmatrix} -12 \\ 5 \end{bmatrix}$$

that are orthogonal because

$$A_1^T A_2 = 5(-12) + 5(12) = 0$$

However, A is not an orthogonal matrix because

$$\det A = 25 + 144 = 169$$

and

$$A^{-1} = \begin{bmatrix} 5/169 & 12/169 \\ -12/169 & 5/169 \end{bmatrix}$$

that does not adhere to the required condition that $A^T = A^{-1}$.

But if

$$A = \begin{bmatrix} 5/13 & -12/13 \\ 12/13 & 5/13 \end{bmatrix}$$

then A_1 and A_2 are

$$A_1 = \begin{bmatrix} 5/13 \\ 12/13 \end{bmatrix} \qquad A_2 = \begin{bmatrix} -12/13 \\ 5/13 \end{bmatrix}$$

and

$$A_1^T A_2 = \frac{1}{169} [5(-12) + 12(5)] = 0$$

which indicates orthogonality. Moreover, with

$$\det A = \frac{1}{169} [5(5) + 12(12)] = 1$$

it is observed that

$$A^T = A^{-1} = \begin{bmatrix} 5/13 & 12/13 \\ -12/13 & 5/13 \end{bmatrix}$$

and A is indeed orthogonal.

The concept of orthogonal matrices applies equally well to $n \times n$ square matrices composed of n, $n \times 1$ column vectors. For an orthonormal set of either column vectors or n, $1 \times n$ row vectors, $A_1, A_2, A_3, \ldots, A_n$, for all i and j

$$A_i^T A_j = A_j^T A_i = \delta_{ij} \tag{6.10}$$

where δ_{ij} is the Kronecker delta function

$$\delta_{ij} = \begin{cases} 0; & i \neq j \\ 1; & i = j \end{cases} \tag{6.11}$$

Equations (6.10) and (6.11) tell it all; orthonormal vectors possess magnitudes equal to unity and vanishing dot or scalar products with each other.

What is more, it is easy to see that if the columns of A form an orthonormal set, so do the rows. Consider A, which is $n \times n$ represented in terms of its row vectors:

$$A = \begin{bmatrix} A_1 \\ A_2 \\ A_3 \\ \cdots \\ A_n \end{bmatrix}$$

The transpose of A must be

$$A^T = [A_1^T \quad A_2^T \quad A_3^T \quad \cdots \quad A_n^T]$$

and if the dot product is taken by means of a postmultiplication of A by A^T,

$$AA^T = \begin{bmatrix} A_1 A_1^T & A_2 A_1^T & A_3 A_1^T & \cdots & A_n A_1^T \\ A_1 A_2^T & A_2 A_2^T & A_3 A_2^T & \cdots & A_n A_2^T \\ A_1 A_3^T & A_2 A_3^T & A_3 A_3^T & \cdots & A_n A_3^T \\ & & \cdots & & \\ A_1 A_n^T & A_2 A_n^T & A_3 A_n^T & \cdots & A_n A_n^T \end{bmatrix}$$

Clearly, by Eqs. (6.10) and (6.11), this is equal to the identity matrix, and

$$A^{T}A = I$$

which shows that the transpose of A is its inverse and that all vectors in A (row or column) form orthonormal sets. This also indicates that the square root of an orthonormal matrix must be equal to $\pm I$ and that the determinant of an orthonormal matrix must be equal to ± 1.

Products of orthogonal matrices must yield orthogonal matrices. If A and B are both orthogonal,

$$(AB)^{T} = B^{T}A^{T} = B^{-1}A^{-1} = (AB)^{-1}$$

This may be extended to any number of orthogonal matrices in cascade:

$$(ABC \cdots YZ)^{T} = (ABC \cdots YZ)^{-1}$$

6.9 THE GRAM–SCHMIDT ORTHOGONALIZATION PROCESS

Orthogonal and orthonormal vectors possess considerable versatility. For example, if an $n \times n$ matrix is nonsingular and orthogonal, its inverse is quickly and handily obtained by merely taking its transpose. Additional conveniences and advantages that pertain to orthogonal and orthonormal vectors become evident as this study proceeds.

A linearly independent set of $n \times 1$ vectors (not necessarily n of them) can be made orthogonal and indeed orthonormal by a procedure known as the *Gram–Schmidt orthogonalization process*. This procedure begins with a set of n linearly independent vectors $A_1, A_2, A_3, \ldots, A_n$, which may or may not be orthonormal. One of them is chosen, say A_1, and the first vector in the orthogonal and orthonormal set is set equal to A_1

$$B_1 = A_1$$

Here, it is presumed that B_1 is not a unit vector. But it is easy to make B_1 a unit vector by dividing it by its length:

$$U_1 = \frac{B_1}{\sqrt{B_1^{T}B_1}}$$

Now take any other of the vectors, say A_2, and make it orthogonal to U_1. The strategy for doing this may be observed in Fig. 6.4, which shows $A_1 = B_1, A_2, U_1$, and $\alpha_1 U_1$ in the Cartesian coordinate system with coordinate axes x_1, x_2, and x_3. Notice here that the vector $B_2 = A_2 - \alpha_1 U_1$ is perpendicular or orthogonal to A_1. The vector B_2 is perpendicular to A_1

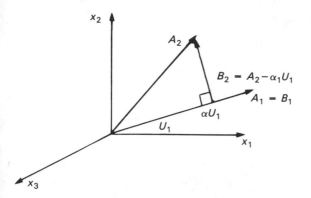

Figure 6.4 Two vectors A_1 and A_2 in the Cartesian coordinate system with coordinate axes x_1, x_2, and x_3.

and U_1 so that

$$U_1^T B_2 = U_1^T (A_2 - \alpha_1 U_2) = U_1^T A_2 - \alpha_1 U_1^T U_1 = 0$$

However, because $U_1^T U_1 = 1$, the constant α_1 can be quickly obtained:

$$\alpha_1 = U_1^T A_2$$

so that

$$B_2 = A_2 - (U_1^T A_2) U_1 \qquad (6.12a)$$

and

$$U_2 = \frac{B_2}{\sqrt{B_2^T B_2}}$$

The next step is to take A_3 and find a vector B_3 that is orthogonal to both U_1 and U_2. Let

$$B_3 = A_3 - \beta_1 U_1 - \beta_2 U_2$$

and premultiply this, first by U_1^T and then by U_2^T,

$$U_1^T B_3 = U_1^T A_3 - \beta_1 U_1^T U_1 - \beta_2 U_1^T U_2$$

$$U_2^T B_3 = U_2^T A_3 - \beta_1 U_2^T U_1 - \beta_2 U_2^T U_2$$

With $U_1^T U_1 = U_2^T U_2 = 1$ and $U_1^T U_2 = U_2^T U_1 = 0$, these become

$$U_1^T B_3 = U_1^T A_3 - \beta_1$$

$$U_2^T B_3 = U_2^T A_3 - \beta_2$$

and because $U_1^T B_3 = U_2^T B_3 = 0$

$$\beta_1 = U_1^T A_3$$

$$\beta_2 = U_2^T A_3$$

so that

$$B_3 = A_3 - (U_1^T A_3)U_1 - (U_2^T A_3)U_2 \qquad (6.12b)$$

and

$$U_3 = \frac{B_3}{\sqrt{B_3^T B_3}}$$

Mathematical induction may now be used to set down the algorithms for performing the Gram–Schmidt orthogonalization process:

$$B_k = A_k - \sum_{i=1}^{k-1} (U_i^T A_k)U_i \qquad (6.13a)$$

and

$$U_k = \frac{B_k}{\sqrt{B_k^T B_k}} \qquad (6.13b)$$

None of the vectors B_k in Eq. (6.13a) can ever be null, because all the vectors A_k are linearly independent.

Example: Suppose it is desired to construct a set of orthonormal vectors from the set

$$A = \begin{bmatrix} 1 \\ 1 \\ 0 \end{bmatrix} \qquad A = \begin{bmatrix} 1 \\ 0 \\ 1 \end{bmatrix} \qquad A = \begin{bmatrix} 0 \\ 1 \\ 1 \end{bmatrix}$$

The first step is to determine whether the given vectors are linearly independent. This can be investigated by means of the Gram determinant:

$$A_1^T A_1 = 2$$

$$A_1^T A_2 = 1$$

$$A_1^T A_3 = 1$$

$$A_2^T A_2 = 2$$

$$A_2^T A_3 = 1$$

and

$$A_3^T A_3 = 2$$

These make the Gram determinant:

$$G = \begin{bmatrix} 2 & 1 & 1 \\ 1 & 2 & 1 \\ 1 & 1 & 2 \end{bmatrix}$$

that has a value, which the reader may verify, det $G = 4$. This says that the given vectors constitute a linearly independent set.

Now let $B_1 = A_1$ so that

$$U_1 = \frac{B_1}{\sqrt{B_1^T B_1}} = \begin{bmatrix} \sqrt{2}/2 \\ \sqrt{2}/2 \\ 0 \end{bmatrix}$$

Now, by Eq. (6.13a),

$$B_2 = A_2 - (U_1^T A_2)U_1$$

and with

$$U_1^T A_2 = \frac{\sqrt{2}}{2} + 0 + 0 = \frac{\sqrt{2}}{2}$$

the vector B_2 is computed as

$$B_2 = \begin{bmatrix} 1 \\ 0 \\ 1 \end{bmatrix} - \frac{\sqrt{2}}{2} \begin{bmatrix} \sqrt{2}/2 \\ \sqrt{2}/2 \\ 0 \end{bmatrix} = \begin{bmatrix} 1 - 1/2 \\ 0 - 1/2 \\ 1 \end{bmatrix} = \begin{bmatrix} 1/2 \\ -1/2 \\ 1 \end{bmatrix}$$

B_2 is not a unit vector. Take

$$B_2^T B_2 = \frac{1}{4} + \frac{1}{4} + 1 = \frac{6}{4}$$

so that

$$U_2 = \frac{2\sqrt{6}}{6} \begin{bmatrix} 1/2 \\ -1/2 \\ 1 \end{bmatrix} = \frac{\sqrt{6}}{6} \begin{bmatrix} 1 \\ -1 \\ 2 \end{bmatrix}$$

The final step involves B_3 and A_3. By Eq. (6.13a),

$$B_3 = A_3 - (U_1^T A_3)U_1 - (U_2^T A_3)U_2$$

and with

$$U_1^T A_3 = 0 + \sqrt{2}/2 + 0 = \sqrt{2}/2$$

and

$$U_2^T A_3 = 0 - \frac{\sqrt{6}}{6} + \frac{2\sqrt{6}}{6} = \frac{\sqrt{6}}{6}$$

B_3 is computed as

$$B_3 = \begin{bmatrix} 0 \\ 1 \\ 1 \end{bmatrix} - \frac{\sqrt{2}}{2}\begin{bmatrix} \sqrt{2}/2 \\ \sqrt{2}/2 \\ 0 \end{bmatrix} - \frac{\sqrt{6}}{6}\begin{bmatrix} \sqrt{6}/6 \\ -\sqrt{6}/6 \\ 2\sqrt{6}/6 \end{bmatrix}$$

$$= \begin{bmatrix} 0 - 1/2 - 1/6 \\ 1 - 1/2 + 1/6 \\ 1 - 0 - 1/3 \end{bmatrix} = \begin{bmatrix} -2/3 \\ 2/3 \\ 2/3 \end{bmatrix}$$

and then with

$$B_3^T B_3 = \frac{4}{9} + \frac{4}{9} + \frac{4}{9} = \frac{12}{9}$$

it is seen that

$$U_3 = \begin{bmatrix} -\sqrt{3}/3 \\ \sqrt{3}/3 \\ \sqrt{3}/3 \end{bmatrix}$$

The reader can easily verify that U_1, U_2, and U_3 form an orthonormal set.

6.10 LINEAR TRANSFORMATIONS

A linear transformation may be defined as a relationship between one set of variables X $(x_1, x_2, x_3, \ldots, x_n)$ and another set of variables Y $(y_1, y_2, y_3, \ldots, y_n)$. If the matrix of the transformation is designated as A, then the linear transformation that maps X into Y is given by

$$Y = AX$$

where A is called the *matrix of the transformation*.

Not all transformations are linear transformations. To be a linear transformation, the transformation must satisfy the homogeneity condition:

$$\alpha Y = A[\alpha X] \tag{6.14a}$$

and the superposition condition:

$$Y_1 + Y_2 = A[X_1 + X_2] \tag{6.14b}$$

Linear transformations may be cascaded. If $Y = AX$ and $Z = BY$, then

$$Z = BY = B(AX) = (BA)X = CX$$

where $C = BA$ is the matrix of the transformation that maps X into Z.

Now if P and Q are nonsingular matrices,

$$B = PAQ$$

represents a linear transformation that maps A into B. Here P can be obtained from the identity matrix by a series of n elementary transformations involving row operations:

$$P = E_{i,n}E_{i,n-1}, \cdots E_{i3}E_{i2}E_{i1}$$

where i may take on the values 1, 2, or 3. The matrix Q can also be obtained from the identity matrix by a series of elementary transformations involving column operations:

$$Q = E_{i1}E_{i2}E_{i3} \cdots E_{i,n-1}E_{i,n}$$

From the definition of the equivalence of matrices (Section 5.9), it is easy to see that both P and Q are equivalent to I. Thus in

$$B = PAQ \tag{6.15}$$

B may be said to be equivalent to A and Eq. (6.15) represents what is called an equivalence transformation.

If P and Q are orthogonal matrices where $Q = P^T = P^{-1}$, then

$$B = PAP^T = PAP^{-1}$$

is called an *orthogonal transformation* and B is said to be orthogonally similar to A.

If P and Q are not orthogonal but $Q = P^T$, then

$$B = PAP^T$$

is designated as a *congruence transformation* with B congruent to A.

Finally, if $Q = P^{-1}$, then

$$B = PAP^{-1}$$

is called a *similarity transformation* with B similar to A. Observe that orthogonal transformations are also congruence and similarity transformations but that the reverse is not necessarily true.

6.11 THE GEOMETRIC INTERPRETATION OF LINEAR TRANSFORMATIONS

In the matrix representation

$$Y = AX$$

the matrix A, which is the matrix of the linear transformation, can be thought of in geometric terms. Depending on the form of A, the mapping of X into Y can be considered as a reflection, a rotation, or a shear.

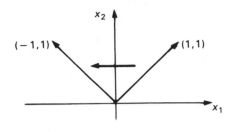

Figure 6.5 A reflection about the vertical axis in the x_1, x_2 coordinate system.

Consider the x_1, x_2 plane in a Cartesian coordinate system and let

$$A = \begin{bmatrix} -1 & 0 \\ 0 & 1 \end{bmatrix}$$

Then in $Y = AX$,

$$\begin{bmatrix} y_1 \\ y_2 \end{bmatrix} = \begin{bmatrix} -1 & 0 \\ 0 & 1 \end{bmatrix}\begin{bmatrix} x_1 \\ x_2 \end{bmatrix} = \begin{bmatrix} -x_1 \\ x_2 \end{bmatrix}$$

and it is observed in Fig. 6.5 that A, in this case, provides a reflection of the vector from the origin to point $(1, 1)$ across the x_2 axis so that the vector terminates at point $(-1, 1)$. It may be said that the point $(1, 1)$ is mapped to the point $(-1, 1)$ by the linear transformation, and it is to be noted that in this kind of reflection (or reflection transformation), the length or magnitude of the vector is preserved.

However, if A is given by

$$A = \begin{bmatrix} 0 & 1 \\ 1 & 0 \end{bmatrix}$$

then

$$\begin{bmatrix} y_1 \\ y_2 \end{bmatrix} = \begin{bmatrix} 0 & 1 \\ 1 & 0 \end{bmatrix}\begin{bmatrix} x_1 \\ x_2 \end{bmatrix} = \begin{bmatrix} x_2 \\ x_1 \end{bmatrix}$$

and, as displayed in Fig. 6.6, this is a reflection across the line whose equation is $x_2 = x_1$. The vector that terminates at the point $(2,1)$ is mapped to a position such that it terminates at the point $(1,2)$. Here, too, it can be noted that the magnitude of the vector is preserved.

But if A is given by the matrix

$$A = \begin{bmatrix} \cos\theta & -\sin\theta \\ \sin\theta & \cos\theta \end{bmatrix}$$

so that

$$\begin{bmatrix} y_1 \\ y_2 \end{bmatrix} = \begin{bmatrix} \cos\theta & -\sin\theta \\ \sin\theta & \cos\theta \end{bmatrix}\begin{bmatrix} x_1 \\ x_2 \end{bmatrix} = \begin{bmatrix} x_1\cos\theta & -x_2\sin\theta \\ x_1\sin\theta & x_2\cos\theta \end{bmatrix}$$

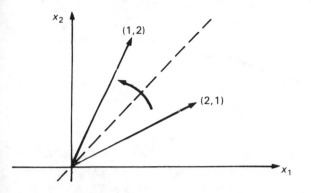

Figure 6.6 Reflection of a vector about an inclined line running through the first quadrant at an angle of 45°.

it is easy to see that, as indicated in Fig. 6.7, this represents a rotation with the center of rotation at the origin. If $\theta = 30°$, the vector that terminates at the point on the x axis $(2, 0)$ is rotated so that its termination point is $(\sqrt{3} = 1.732, 1.000)$. The mapping here is from the point $(2, 0)$ to the point $(1.732, 1.000)$. In this case, the magnitude of the vector is preserved.

A linear transformation of the form

$$A = \begin{bmatrix} 1 & \alpha \\ 0 & 1 \end{bmatrix}$$

provides what is called a *shear*. Figure 6.8 demonstrates that in

$$\begin{bmatrix} y_1 \\ y_2 \end{bmatrix} = \begin{bmatrix} 1 & \alpha \\ 0 & 1 \end{bmatrix} \begin{bmatrix} x_1 \\ x_2 \end{bmatrix} = \begin{bmatrix} x_1 + \alpha x_2 \\ x_2 \end{bmatrix}$$

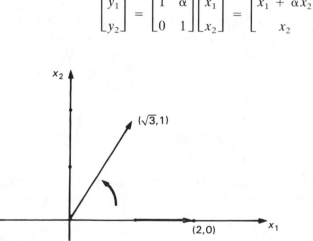

Figure 6.7 Rotation of a vector counterclockwise with the axis of rotation through the origin and perpendicular to the x_1, x_2 plane.

Figure 6.8 In the shear suggested by the horizontal lines, the horizontal or x_1 axis is fixed and the vector is shifted to the right. Each horizontal line is shifted a distance proportional to its distance above the x_1 axis.

with $\alpha = 2$, the vector that terminates at the point $(1, 1)$ is shifted so that its termination point is $(3, 1)$. The transformation maps the point $(1, 1)$ to $(3, 1)$, and the magnitude of the vector is not preserved. The idea of a shear is suggested by the horizontal lines which resemble the side view of a deck of playing cards.

This brief summary pertaining to the geometric interpretations of linear transformations is interesting and clearly shows how the transformations involve a movement of vectors or a mapping of one point to another. Of possibly greater importance is the representation of a vector in various coordinate systems. This requires what is called a *transformation of coordinates* and is the subject of the next section.

6.12 THE TRANSFORMATION OF COORDINATES

Consider Fig. 6.9 which shows a vector P in a Cartesian coordinate system x_1, x_2, and x_3. This system may be referred to as the *x-coordinate system*. Another Cartesian coordinate system, y_1, y_2, and y_3, is also shown, and it should be noted that the origins of the two coordinate systems coincide. The vector P, of course, can also be considered to exist in the y-coordinate system.

The vector P can be represented in the x-coordinate system as the vector X in terms of the coordinates and the unit vectors **i**, **j**, and **k**

$$X = x_1\mathbf{i} + x_2\mathbf{j} + x_3\mathbf{k} \tag{6.16a}$$

If the unit vectors in the y-coordinate system are designated as **i′**, **j′**, and **k**-, then the vector P designated by Y in the y-coordinate system can be set down as

$$Y = y_1\mathbf{i'} + y_2\mathbf{j'} + y_3\mathbf{k'} \tag{6.16b}$$

It must be clearly understood that the foregoing proposals for the repre-

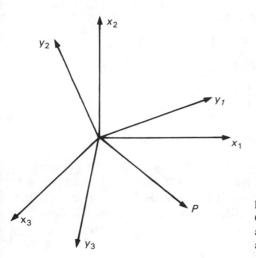

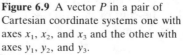

Figure 6.9 A vector P in a pair of Cartesian coordinate systems one with axes x_1, x_2, and x_3 and the other with axes y_1, y_2, and y_3.

sentation of P in both coordinate systems do not violate the premise that

$$P = X = Y$$

or

$$P = x_1\mathbf{i} + x_2\mathbf{j} + x_3\mathbf{k} = y_1\mathbf{i}' + y_2\mathbf{j}' + y_3\mathbf{k}' \qquad (6.17)$$

If scalar or dot products of Eqs. (6.16) are taken in turn with the unit vectors in the x-coordinate system $\mathbf{i}$, $\mathbf{j}$, and $\mathbf{k}$, the result is

$$X \cdot \mathbf{i} = x_1\mathbf{i} \cdot \mathbf{i} \quad + x_2\mathbf{j} \cdot \mathbf{i} \quad + x_3\mathbf{k} \cdot \mathbf{i}$$

$$X \cdot \mathbf{j} = x_1\mathbf{i} \cdot \mathbf{j} \quad + x_2\mathbf{j} \cdot \mathbf{j} \quad + x_3\mathbf{k} \cdot \mathbf{j}$$

$$X \cdot \mathbf{k} = x_1\mathbf{i} \cdot \mathbf{k} \quad + x_2\mathbf{j} \cdot \mathbf{k} \quad + x_3\mathbf{k} \cdot \mathbf{k}$$

and

$$Y \cdot \mathbf{i} = y_1\mathbf{i}' \cdot \mathbf{i} + y_2\mathbf{j}' \cdot \mathbf{i} + y_3\mathbf{k}' \cdot \mathbf{i}$$

$$Y \cdot \mathbf{j} = y_1\mathbf{i}' \cdot \mathbf{j} + y_2\mathbf{j}' \cdot \mathbf{j} + y_3\mathbf{k}' \cdot \mathbf{j}$$

$$Y \cdot \mathbf{k} = y_1\mathbf{i}' \cdot \mathbf{k} + y_2\mathbf{j}' \cdot \mathbf{k} + y_3\mathbf{k}' \cdot \mathbf{k}$$

But because $\mathbf{i}$, $\mathbf{j}$, and $\mathbf{k}$ are orthogonal unit vectors, $\mathbf{i} \cdot \mathbf{i} = \mathbf{j} \cdot \mathbf{j} = \mathbf{k} \cdot \mathbf{k} = 1$ and $\mathbf{i} \cdot \mathbf{j} = \mathbf{j} \cdot \mathbf{k} = \mathbf{k} \cdot \mathbf{i} = 0$. Thus, as demanded by Eq. (6.17),

$$x_1 = y_1\mathbf{i}' \cdot \mathbf{i} + y_2\mathbf{j}' \cdot \mathbf{i} + y_3\mathbf{k}' \cdot \mathbf{i}$$

$$x_2 = y_1\mathbf{i}' \cdot \mathbf{j} + y_2\mathbf{j}' \cdot \mathbf{j} + y_3\mathbf{k}' \cdot \mathbf{j}$$

$$x_3 = y_1\mathbf{i}' \cdot \mathbf{k} + y_2\mathbf{j}' \cdot \mathbf{k} + y_3\mathbf{k}' \cdot \mathbf{k}$$

This may be more compactly written as

$$X = TY \tag{6.18}$$

where T is a transformation matrix given by

$$T = \begin{bmatrix} t_{11} & t_{12} & t_{13} \\ t_{21} & t_{22} & t_{23} \\ t_{31} & t_{32} & t_{33} \end{bmatrix} = \begin{bmatrix} \mathbf{i}' \cdot \mathbf{i} & \mathbf{j}' \cdot \mathbf{i} & \mathbf{k}' \cdot \mathbf{i} \\ \mathbf{i}' \cdot \mathbf{j} & \mathbf{j}' \cdot \mathbf{j} & \mathbf{k}' \cdot \mathbf{j} \\ \mathbf{i}' \cdot \mathbf{k} & \mathbf{j}' \cdot \mathbf{k} & \mathbf{k}' \cdot \mathbf{k} \end{bmatrix} \tag{6.19}$$

The elements of the transformation matrix are just exactly the direction cosines of the unit vectors $\mathbf{i}'$, $\mathbf{j}'$, and $\mathbf{k}'$ in the y-coordinate system with respect to the x-coordinate system. It has already been noted (Sec. 6.6) that the sum of the squares of the direction cosines for any coordinate axis must be unity. Thus

$$\mathbf{i}' \cdot \mathbf{i} + \mathbf{j}' \cdot \mathbf{i} + \mathbf{k}' \cdot \mathbf{i} = 1$$

$$\mathbf{i}' \cdot \mathbf{j} + \mathbf{j}' \cdot \mathbf{j} + \mathbf{k}' \cdot \mathbf{j} = 1$$

$$\mathbf{i}' \cdot \mathbf{k} + \mathbf{j}' \cdot \mathbf{k} + \mathbf{k}' \cdot \mathbf{k} = 1$$

and each row and column vector in the matrix T has a magnitude of unity. Now consider the matrix T to be composed of three column vectors:

$$T = [T_1 \quad T_2 \quad T_3]$$

It is of great interest as to whether these three unit column vectors form an orthogonal set. An investigation of the orthogonality of these column vectors requires a short digression into the realm of vector cross products and scalar triple products.

Consider two vectors A and B and define their cross product as

$$C = A \times B = |A||B|\sin \alpha \, \mathbf{n}$$

where C is a a vector in the direction of n that is normal to the plane containing A and B. The angle α is the smaller of the two angles between A and B, and the positive sense of the unit vector $\mathbf{n}$ is in the direction of a right-hand screw thread as A is rotated into B through the angle α. Observe that this right-hand rule means that

$$A \times B = -B \times A$$

Example: If

$$A = a_1\mathbf{i} + a_2\mathbf{j} + a_3\mathbf{k}$$

and

$$B = b_1\mathbf{i} + b_2\mathbf{j} + b_3\mathbf{k}$$

the cross product $A \times B$ is

$$A \times B = (a_1\mathbf{i} + a_2\mathbf{j} + a_3\mathbf{k}) \times (b_1\mathbf{i} + b_2\mathbf{j} + b_3\mathbf{k})$$
$$= a_1b_1\mathbf{i} \times \mathbf{i} + a_1b_2\mathbf{i} \times \mathbf{j} + a_1b_3\mathbf{i} \times \mathbf{k}$$
$$+ a_2b_1\mathbf{j} \times \mathbf{i} + a_2b_2\mathbf{j} \times \mathbf{j} + a_2b_3\mathbf{j} \times \mathbf{k}$$
$$+ a_3b_1\mathbf{k} \times \mathbf{i} + a_3b_2\mathbf{k} \times \mathbf{j} + a_3b_3\mathbf{k} \times \mathbf{k}$$

But the unit vectors i, j, and k are orthogonal so that

$$\mathbf{i} \times \mathbf{i} = \mathbf{j} \times \mathbf{j} = \mathbf{k} \times \mathbf{k} = 0$$
$$\mathbf{i} \times \mathbf{j} = \mathbf{j} \times \mathbf{k} = \mathbf{k} \times \mathbf{i} = 1$$

and

$$\mathbf{j} \times \mathbf{i} = \mathbf{i} \times \mathbf{k} = \mathbf{k} \times \mathbf{j} = -1$$

Hence

$$A \times B = a_1b_2\mathbf{k} - a_1b_3\mathbf{j} - a_2b_1\mathbf{k} + a_2b_3\mathbf{i} + a_3b_1\mathbf{j} - a_3b_2\mathbf{i}$$

or

$$A \times B = (a_2b_3 - a_3b_2)\mathbf{i} - (a_1b_3 - a_3b_1)\mathbf{j} + (a_1b_2 - a_2b_1)\mathbf{k}$$

The result of the example shows that the cross product may be written quite simply and handily as a determinant:

$$A \times B = \begin{vmatrix} \mathbf{i} & \mathbf{j} & \mathbf{k} \\ a_1 & a_2 & a_3 \\ b_1 & b_2 & b_3 \end{vmatrix}$$

The *scalar triple product* (also called the *triple scalar product*) of the three vectors A, B, and C is defined as

$$A \cdot B \times C = |A, B, C| = \det[A, B, C]$$

The matrix $[A, B, C]$ is given by

$$[A,B,C] = \begin{vmatrix} a_1 & a_2 & a_3 \\ b_1 & b_2 & b_3 \\ c_1 & c_2 & c_3 \end{vmatrix}$$

as the next example will clearly show.

Example: With the vectors A, B, and C written in terms of their coordinates and unit vectors, it is easy to see that $B \times C$ is a vector

$$B \times C = \begin{vmatrix} \mathbf{i} & \mathbf{j} & \mathbf{k} \\ b_1 & b_2 & b_3 \\ c_1 & c_2 & c_3 \end{vmatrix}$$

that is equal to

$$D = B \times C = (b_2 c_3 - b_3 c_2)\mathbf{i} + (b_3 c_1 - b_1 c_3)\mathbf{j} + (b_1 c_2 - b_2 c_1)\mathbf{k}$$

or

$$D = B \times C = (b_2 c_3 - b_3 c_2)\mathbf{i} - (b_1 c_3 - b_3 c_1)\mathbf{j} + (b_1 c_2 - b_2 c_1)\mathbf{k}$$

The dot or scalar product may now be taken:

$$A = (a_1\mathbf{i} + a_2\mathbf{j} + a_3\mathbf{k}) \cdot D$$

or

$$A \cdot D = a_1(b_2 c_3 - b_3 c_2) - a_2(b_1 c_3 - b_3 c_1) + a_3(b_1 c_2 - b_2 c_1)$$

and it is observed that this expression is nothing more than the cofactor expansion of the determinant

$$\det[A, B, C] = \begin{vmatrix} a_1 & a_2 & a_3 \\ b_1 & b_2 & b_3 \\ c_1 & c_2 & c_3 \end{vmatrix}$$

along its upper row.

Now T, which is square and nonsingular, possesses an inverse T^{-1}. Let T^{-1} be composed of the row vectors E, F, and G so that

$$TT^{-1} = [T_1 \quad T_2 \quad T_3] \begin{bmatrix} E \\ F \\ G \end{bmatrix} = I$$

This requires that

$$T_1 \cdot E = T_2 \cdot F = T_3 \cdot G = 1$$

and

$$T_1 \cdot F = T_1 \cdot G = T_2 \cdot E = T_2 \cdot G = T_3 \cdot E = T_3 \cdot F = 0$$

Clearly, E must be perpendicular to both T_2 and T_3, F must be perpendicular to T_1 and T_3, and G must be perpendicular to T_1 and T_2.

This point is made by working with E, and the conclusions that pertain to F and G then follow logically. Invoke the cross product

$$E = \alpha T_2 \times T_3$$

With this in $T_1 \cdot E = 1$, it is seen that

$$T_1 \cdot (\alpha T_2 \times T_3) = \alpha[T_1, T_2, T_3] = 1$$

and if $\det[T_1, T_2, T_3] \neq 0$, then

$$\alpha = \frac{1}{\det[T_1, T_2, T_3]}$$

and

$$E = \frac{T_2 \times T_3}{\det[T_1, T_2, T_3]}$$

It can be shown in a similar development that

$$F = \frac{T_3 \times T_1}{\det[T_1, T_2, T_3]}$$

and

$$G = \frac{T_1 \times T_2}{\det[T_1, T_2, T_3]}$$

All of this means that the inverse of T can be written as

$$T^{-1} = \frac{1}{\det[T_1, T_2, T_3]} \begin{bmatrix} T_2 \times T_3 \\ T_3 \times T_1 \\ T_1 \times T_2 \end{bmatrix}$$

and it is said the vectors T_1, T_2, and T_3 are self-reciprocal to the vectors E, F, and G.

Return now to the transformation matrix T:

$$T = \begin{bmatrix} \mathbf{i}' \cdot \mathbf{i} & \mathbf{j}' \cdot \mathbf{i} & \mathbf{k}' \cdot \mathbf{i} \\ \mathbf{i}' \cdot \mathbf{j} & \mathbf{j}' \cdot \mathbf{j} & \mathbf{k}' \cdot \mathbf{j} \\ \mathbf{i}' \cdot \mathbf{k} & \mathbf{j}' \cdot \mathbf{k} & \mathbf{k}' \cdot \mathbf{k} \end{bmatrix}$$

It is easy to see that $\mathbf{i}$, $\mathbf{j}$, and $\mathbf{k}$ are self-reciprocal to the vectors $\mathbf{i}$, $\mathbf{j}$, and $\mathbf{k}$. Moreover, the columns of T represent the unit vectors $\mathbf{i}$, $\mathbf{j}$, and $\mathbf{k}$ in the x-coordinate system (the $\mathbf{i}$, $\mathbf{j}$, $\mathbf{k}$ system). Because of the self-reciprocal feature, one may write

$$T^{-1} = \begin{bmatrix} \mathbf{i}' \cdot \mathbf{i} & \mathbf{i}' \cdot \mathbf{j} & \mathbf{i}' \cdot \mathbf{k} \\ \mathbf{j}' \cdot \mathbf{i} & \mathbf{j}' \cdot \mathbf{j} & \mathbf{j}' \cdot \mathbf{k} \\ \mathbf{k}' \cdot \mathbf{i} & \mathbf{k}' \cdot \mathbf{j} & \mathbf{k}' \cdot \mathbf{k} \end{bmatrix} = T^{\mathrm{T}}$$

and there it is; the transformation matrix is orthogonal and indeed orthonormal.

This is why the transformation of coordinates is referred to as a *linear orthogonal transformation*. It is a transformation between two Cartesian coordinate systems that have the same origin:

$$Y = TX$$

and

$$X = T^{-1}Y$$

Moreover, if X is mapped or transformed into Y by T_1 and Y in turn can be transformed into Z by T_2, then

$$Y = T_1 X$$

$$Z = T_2 Y$$

and

$$Z = T_2 Y = T_2(T_1 X) = (T_2 T_1)X = T_3 X$$

where $T_3 = T_2 T_1$. Of course, one may go back to X from Z via

$$X = T_3^{-1} Z$$

Now suppose that it is desired to rotate a Cartesian coordinate system, designated as the x system, with axes x_1, x_2, and x_3 around the x_3 axis by an angle θ. This yields another Cartesian coordinate system that can be designated as the y system with axes y_1, y_2, and y_3, where the y_3 axis is coincident with the x_3 axis. Such a rotation is displayed in Fig. 6.10 and it is observed that the direction cosines are:

$$\text{from } x_1 \text{ to } y_1; \cos \theta$$

$$\text{from } x_1 \text{ to } y_2; \cos \left(\frac{\pi}{2} + \theta \right) = -\sin \theta$$

$$\text{from } x_1 \text{ to } y_3; \cos \frac{\pi}{2} = 0$$

$$\text{from } x_2 \text{ to } y_1; \cos \left(\frac{\pi}{2} - \theta \right) = \sin \theta$$

$$\text{from } x_2 \text{ to } y_2; \cos \theta$$

$$\text{from } x_2 \text{ to } y_3; \cos \frac{\pi}{2} = 0$$

$$\text{from } x_3 \text{ to } y_1; \cos \frac{\pi}{2} = 0$$

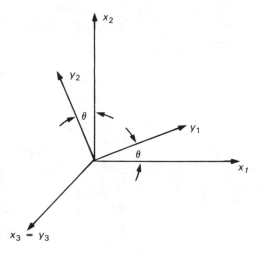

Figure 6.10 Two Cartesian coordinate systems with common origin. The y_1 and y_2 axes are rotated about the $y_3 = x_3$ axes from the x_1 and x_2 axes by an angle θ.

from x_3 to y_2; $\cos \dfrac{\pi}{2} = 0$

from x_3 to y_3; $\cos 0 = 1$

This makes the transformation matrix T in $Y = TX$

$$T = \begin{bmatrix} \cos\theta & \sin\theta & 0 \\ -\sin\theta & \cos\theta & 0 \\ 0 & 0 & 1 \end{bmatrix} \tag{6.20}$$

and it can be noted that both the row and column vectors in T are indeed orthonormal.

Example: If the angle θ in the transformation matrix shown in Eq. (6.20) is prescribed to be 30°, then

$$T = \begin{bmatrix} 0.8660 & 0.5000 & 0 \\ -0.5000 & 0.8660 & 0 \\ 0 & 0 & 1 \end{bmatrix}$$

and this indicates that a vector in a y-coordinate system can be obtained from the vector

$$X = \begin{bmatrix} 1 \\ 2 \\ 3 \end{bmatrix}$$

in the x-coordinate system by the simple expedient of a matrix multiplication, $Y = TX$. Moreover, the form of the transformation T indicates that this transformation is a rotation around the $x_3 = y_3$ axes. Thus

$$Y = \begin{bmatrix} 0.8660 & 0.5000 & 0 \\ -0.5000 & 0.8660 & 0 \\ 0 & 0 & 1 \end{bmatrix} \begin{bmatrix} 1.0000 \\ 2.0000 \\ 3.0000 \end{bmatrix} = \begin{bmatrix} 1.8660 \\ 1.2321 \\ 3.0000 \end{bmatrix}$$

Because T is an orthonormal matrix $T^{-1} = T^{\mathrm{T}}$ and X can be retrieved from Y by a simple computation involving the inverse of T,

$$X = \begin{bmatrix} 0.8660 & -0.5000 & 0 \\ 0.5000 & 0.8660 & 0 \\ 0 & 0 & 1 \end{bmatrix} \begin{bmatrix} 1.8660 \\ 1.2321 \\ 3.0000 \end{bmatrix} = \begin{bmatrix} 1.0000 \\ 2.0000 \\ 3.0000 \end{bmatrix}$$

Example: Now consider the rotation of X into Y as in the previous example and further consider an additional rotation of Y into Z in a z-coordinate system by a rotation of $45°$ about the $y_3 = z_3$ axis. In this case, T_1 takes X into Y:

$$T_1 = \begin{bmatrix} 0.8660 & 0.5000 & 0 \\ -0.5000 & 0.8660 & 0 \\ 0 & 0 & 1 \end{bmatrix}$$

and a new transformation matrix T_2 that takes Y into Z is established:

$$T_2 = \begin{bmatrix} 0.7071 & 0.7071 & 0 \\ -0.7071 & 0.7071 & 0 \\ 0 & 0 & 1 \end{bmatrix}$$

Hence, to go from Y to Z,

$$Z = \begin{bmatrix} 0.7071 & 0.7071 & 0 \\ -0.7071 & 0.7071 & 0 \\ 0 & 0 & 1 \end{bmatrix} \begin{bmatrix} 1.8660 \\ 1.2321 \\ 3.0000 \end{bmatrix} = \begin{bmatrix} 2.1907 \\ -0.4483 \\ 3.0000 \end{bmatrix}$$

Then with $T_1^{-1} = T_1^{\mathrm{T}}$, Y can be reclaimed:

$$Y = \begin{bmatrix} 0.7071 & -0.7071 & 0 \\ 0.7071 & 0.7071 & 0 \\ 0 & 0 & 1 \end{bmatrix} \begin{bmatrix} 2.1907 \\ -0.4483 \\ 3.0000 \end{bmatrix} = \begin{bmatrix} 1.8660 \\ 1.2321 \\ 3.0000 \end{bmatrix}$$

Moreover, one may go directly from X to Z without ever considering Y by forming $T_3 = T_2 T_1$.

$$T_3 = \begin{bmatrix} 0.7071 & 0.7071 & 0 \\ -0.7071 & 0.7071 & 0 \\ 0 & 0 & 1 \end{bmatrix} \begin{bmatrix} 0.8660 & 0.5000 & 0 \\ -0.5000 & 0.8660 & 0 \\ 0 & 0 & 1 \end{bmatrix}$$

$$= \begin{bmatrix} 0.2588 & 0.9659 & 0 \\ -0.9659 & 0.2588 & 0 \\ 0 & 0 & 1 \end{bmatrix}$$

Then

$$Z = \begin{bmatrix} 0.2588 & 0.9659 & 0 \\ -0.9659 & 0.2588 & 0 \\ 0 & 0 & 1 \end{bmatrix} \begin{bmatrix} 1.0000 \\ 2.0000 \\ 3.0000 \end{bmatrix} = \begin{bmatrix} 2.1907 \\ -0.4483 \\ 3.0000 \end{bmatrix}$$

and with $T_3^{-1} = T_3^T$, one may return to X:

$$X = \begin{bmatrix} 0.2588 & -0.9659 & 0 \\ 0.9659 & 0.2588 & 0 \\ 0 & 0 & 1 \end{bmatrix} \begin{bmatrix} 2.1907 \\ -0.4483 \\ 3.0000 \end{bmatrix} = \begin{bmatrix} 1.0000 \\ 2.0000 \\ 3.0000 \end{bmatrix}$$

Notice that to transform from X to Z, the angle must be $30° + 45° = 75°$. Everything checks, because $\cos 75° = 0.2588$ and $\sin 75° = 0.9659$.

6.13 EXERCISES

Given the vectors

$$A = [2 \quad -1 \quad 3 \quad 0 \quad 4]$$
$$B = [0 \quad 0 \quad 1 \quad 2 \quad 3]$$
$$C = [x \quad 0 \quad 1 \quad -2 \quad 3]$$

and

$$D = [1 \quad y \quad 0 \quad -1 \quad -1]$$

6.1 Determine $A \cdot B$.

6.2 Find the value of x so that C is orthogonal to A.

6.3 Find the value of y so that D is orthogonal to A.

6.4 If $x = 4$, what is the magnitude of C?

6.5 If $y = 2$, what is the magnitude of D?

6.6 If $x = 2$ and $y = -2$, determine whether A, B, C, and D form a linearly independent set.

6.7 Normalize A (convert it to a unit vector).

6.8 Normalize B.

6.9 What is the angle between A and B?

6.10 What is the distance between A and B?

6.11 What are the five direction cosines of A?

6.12 What are the five direction cosines of B?

6.13 What is the dot or scalar product of $2A$ and $3B$?

6.14 Convert the matrix A to an orthonormal matrix:

$$A = \begin{bmatrix} 1 & -2 & 0 \\ 0 & 1 & 3 \\ 1 & 4 & -1 \end{bmatrix}$$

6.15 Consider the vector X in the Cartesian coordinate system, x_1, x_2, and x_3. What transformation matrix is necessary to rotate the coordinates counterclockwise to a new Cartesian coordinate system y_1, y_2, and $y_3 = x_3$ (a rotation about the x_3 coordinate) if the angle of rotation is to be 20°?

6.16 If in exercise 6.15,

$$X = \begin{bmatrix} 3 \\ 4 \\ 5 \end{bmatrix}$$

what is the value of Y?

6.17 Suppose another Cartesian coordinate system z_1, z_2, and $z_3 = x_3$ is obtained from the x-coordinate system of exercise 6.15 by a clockwise rotation about the x_3 axis of 40°. If the vector X is given as in exercise 6.16, what is the value of Z?

6.18 With the value of Z as obtained in exercise 6.17, use the transformation matrices of exercises 6.15 and 6.17 to determine the value of the vector Y.

SEVEN

THE EIGENVALUE PROBLEM

7.1 INTRODUCTION

Matrix equations of the form

$$AX = \lambda X \tag{7.1}$$

occur frequently in engineering analyses. Consider, for example, that the variables of interest in the transient analysis of a linear system are x_1, x_2, and x_3 and that they are related by three linear, simultaneous, nonhomogeneous differential equations with constant coefficients:

$$\frac{dx_1}{dt} + 3x_1 - 2x_2 - 4x_3 = 12$$

$$-x_1 + \frac{dx_2}{dt} + 2x_2 = 0$$

$$-2x_1 + x_2 + \frac{dx_3}{dt} = 0$$

These may be solved for the derivatives

$$\frac{dx_1}{dt} = -3x_1 + 2x_2 + 4x_3 + 12$$

$$\frac{dx_2}{dt} = x_1 - 2x_2$$

$$\frac{dx_3}{dt} = 2x_1 - x_2$$

and then put into matrix form

$$\begin{bmatrix} \dfrac{dx_1}{dt} \\[2ex] \dfrac{dx_2}{dt} \\[2ex] \dfrac{dx_3}{dt} \end{bmatrix} = \begin{bmatrix} -3 & 2 & 4 \\ 1 & -2 & 0 \\ 2 & -1 & 0 \end{bmatrix} \begin{bmatrix} x_1 \\ x_2 \\ x_3 \end{bmatrix} + \begin{bmatrix} 12 \\ 0 \\ 0 \end{bmatrix}$$

and this may be written with $\dot{X}$ indicating the derivative with respect to time as

$$\dot{X} = AX + B \tag{7.2}$$

where

$$A = \begin{bmatrix} -3 & 2 & 4 \\ 1 & -2 & 0 \\ 2 & -1 & 0 \end{bmatrix}$$

and where B represents the column vector of "forcing functions,"

$$\begin{bmatrix} 12 \\ 0 \\ 0 \end{bmatrix}$$

The solution to the system of differential equations begins with the determination of the so-called complementary function. The procedure is to make the set of equations homogeneous and then, knowing that exponential solutions exist, assume that the complementary function is in the form $x = \gamma e^{\lambda t}$ where γ is an arbitrary constant. Thus in

$$\dot{X}_c = AX_c$$

one takes $X_c = Ce^{\lambda t}$ where C is a 3×1 column vector of arbitrary constants. Then with $\dot{X}_c = \lambda Ce^{\lambda t}$, it is easy to see that

$$\lambda X_c = AX_c$$

which is in the form of Eq. (7.1) and where the values of the λs must then be determined.

This describes what is called the *eigenvalue* or *characteristic value problem*. It occurs frequently in engineering analysis in all disciplines. It would be a mistake, however, to believe that the eigenvalue problem comes forth only from a set of differential equations.

Consider the set of linear, simultaneous algebraic equations

$$AX = Y$$

in which the column vector of constants may have been derived from a linear transformation:

$$Y = \lambda X$$

Then

$$A\dot{X} = Y = \lambda X$$

and this is in the form of Eq. (7.1):

$$AX = \lambda X$$

and the scalar multiplier λ must be determined.

7.2 EIGENVALUES

Equation (7.1) may be rearranged to the form

$$[A - \lambda I]X = 0 \tag{7.3}$$

Here the matrix

$$K = [A - \lambda I] \tag{7.4}$$

is called the *characteristic matrix* of the matrix A and for the homogeneous set of equations represented by Eq. (7.3), nontrivial solutions can only be obtained† if

$$\det K = \det [A - \lambda I] = 0 \tag{7.5}$$

The determinant of K is called the *characteristic determinant* or the *characteristic function* of the matrix A, and it is clearly a polynomial in λ:

$$p(\lambda) = \alpha_n \lambda^n + \alpha_{n-1} \lambda^{n-1} + \alpha_{n-2} \lambda^{n-2} + \cdots + \alpha_1 \lambda + \alpha_0 \tag{7.6}$$

This polynomial is called the *characteristic polynomial* of the matrix A, and when it is set equal to zero in accordance with Eq. (7.5),

$$p(\lambda) = \alpha_n \lambda_n + \alpha_{n-1} \lambda^{n-1} + \alpha_{n-2} \lambda^{n-2} + \cdots + \alpha_1 \lambda + \alpha_0 = 0 \tag{7.7}$$

it is called the *characteristic equation* of the matrix A. The λs are the roots of $p(\lambda) = 0$, and because this equation is called the characteristic equation, the roots are called the *characteristic values* or *eigenvalues*. Although in

†See Section 5.8.

this book only real eigenvalues are considered, the eigenvalues are, in general, complex numbers and, of course, some may be repeated.

Example: Consider the matrix A:

$$A = \begin{bmatrix} 0 & 1 & 0 \\ 0 & 0 & 1 \\ -6 & -11 & -6 \end{bmatrix}$$

and form the characteristic matrix $K = [A - \lambda I]$:

$$K = [A - \lambda I] = \begin{bmatrix} -\lambda & 1 & 0 \\ 0 & -\lambda & 1 \\ -6 & -11 & -(6 + \lambda) \end{bmatrix}$$

The evaluation of the determinant of the characteristic matrix yields the characteristic polynomial and the characteristic equation

$$p(\lambda) = -\lambda^3 - 6\lambda^2 - 11\lambda - 6 = 0$$

or

$$p(\lambda) = \lambda^3 + 6\lambda^2 + 11\lambda + 6 = 0$$

The roots of $p(\lambda) = 0$ are easy to find, and these are the eigenvalues that may be listed in order with the largest first:

$$\lambda_1 = -1$$

$$\lambda_2 = -2$$

and

$$\lambda_3 = -3$$

These eigenvalues are real, separate, and distinct.

7.3 PROPERTIES OF THE CHARACTERISTIC POLYNOMIAL

The characteristic equation of an $n \times n$ square matrix A can be written as

$$p(\lambda) = \det [A - \lambda I] = (-1)^n(\alpha_n\lambda^n + \alpha_{n-1}\lambda^{n-1}$$

$$+ \alpha_{n-2}\lambda^{n-2} + \cdots + \alpha_1\lambda + \alpha_0) = 0 \quad (7.8a)$$

If the equation has n distinct roots that are the characteristic or eigenvalues of A, the polynomial may be written in factored form:

$$p(\lambda) = (\lambda - \lambda_1)(\lambda - \lambda_2)(\lambda - \lambda_3) \cdots (\lambda - \lambda_n) = 0 \quad (7.8b)$$

If λ is set equal to zero, these two representations provide

$$\det A = (-1)^n \alpha_0 = \lambda_1 \lambda_2 \lambda_3 \cdots \lambda_n$$

which shows that the product of the eigenvalues is equal to the determinant of the matrix A. Indeed, if A is nonsingular, A cannot possess a zero eigenvalue, and conversely, if just one eigenvalue of A is zero, then A must be singular.

If $\det[A - \lambda I]$ is expanded or if the factors of Eq. (7.8b) are multiplied,

$$p(\lambda) = \cdots \lambda^4 - (\lambda_1 + \lambda_2 + \lambda_3 + \lambda_4 + \cdots)\lambda^3 +$$

$$(\lambda_1\lambda_2 + \lambda_1\lambda_3 + \lambda_1\lambda_4 + \lambda_2\lambda_3 + \lambda_2\lambda_4 + \lambda_3\lambda_4 + \cdots)\lambda^2 -$$

$$(\lambda_1\lambda_2\lambda_3 + \lambda_1\lambda_2\lambda_4 + \lambda_1\lambda_3\lambda_4 + \lambda_2\lambda_3\lambda_4 + \cdots)\lambda + \cdots = 0$$

and a comparison to Eq. (7.8a) indicates that α_{n-1}, the coefficient of the $(n - 1)$th term, is

$$\alpha_{n-1} = -(\lambda_1 + \lambda_2 + \lambda_3 + \lambda_4 + \cdots)$$

An expansion of $\det[A - \lambda I]$ can also be conducted to show that the coefficient of the $(n - 1)$th term is

$$\alpha_{n-1} = -(a_{11} + a_{22} + a_{33} + a_{44} + \cdots)$$

The comparison of these two expressions for α_{n-1} shows that the sum of the eigenvalues of A is equal to the trace of A.

7.4 METHODS FOR THE GENERATION OF THE CHARACTERISTIC POLYNOMIAL

The characteristic polynomial may be generated in many ways, the most obvious of which is the direct expansion of $\det[A - \lambda I] = 0$. This has been demonstrated in the example in Section 7.2.

Three other methods are considered here, the method of principal minors, Bocher's method, and Leverrier's method which also can be used to provide the inverse of A.

7.5 THE CHARACTERISTIC POLYNOMIAL FROM THE PRINCIPAL MINORS OF A MATRIX

The idea of a principal submatrix of a square matrix has been presented in Section 5.14, where it was pointed out that a principal submatrix of A is any submatrix of A whose principal diagonal is part of the principal diagonal of A.

The expansion of the characteristic equation in Section 7.3 shows how the coefficients of $p(\lambda)$ involves sums of eigenvalues, sums of eigenvalues taken two at a time, sums of eigenvalues taken three at a time, on up to the product of all the eigenvalues. It is not hard to see that, in view of this, the characteristic polynomial can be formed from a consideration of the principal minors of A that are the determinants of the principal submatrices of A. A nice little rule provides a prescription for this.

The coefficient that is associated with the kth term in the characteristic polynomial $p(\lambda)$ of the nth order matrix A is $(-1)^k$ times the sum of the principal minors of order $n - k$ of A. Moreover, the coefficient of λ^n is $(-1)^n$ and the constant term is equal to det A.

Example: The matrix A:

$$A = \begin{bmatrix} 0 & 1 & 0 \\ 0 & 0 & 1 \\ -6 & -11 & -6 \end{bmatrix}$$

has three first-order principal minors given by the elements along the principal diagonal

$$0, \quad 0, \quad -6$$

three second-order principal minors

$$\begin{vmatrix} 0 & 1 \\ 0 & 0 \end{vmatrix} = 0, \qquad \begin{vmatrix} 0 & 1 \\ -11 & -6 \end{vmatrix} = 11, \qquad \begin{vmatrix} 0 & 0 \\ -6 & -6 \end{vmatrix} = 0$$

and the principal minor which is the determinant of A, det $A = -6$
The characteristic equation is

$$p(\lambda) = (-1)^3\lambda^3 + (-1)^2(-6)\lambda^2 + (-1)^1(11)\lambda + (-6) = 0$$

or

$$p(\lambda) = -\lambda^3 - 6\lambda^2 - 11\lambda - 6 = \lambda^3 + 6\lambda^2 + 11\lambda + 6 = 0$$

7.6 BOCHER'S METHOD†

Bocher's method for the generation of the characteristic polynomial $p(\lambda)$ of the $n \times n$ matrix A involves the trace‡ of various powers of A.

---

†Bocher, M., *Introduction to Higher Algebra*, Dover Publications, New York, 1954.

‡Some authors refer to the trace of A (tr A) as the spur of A (sp A).

The first step is to compute n powers of A, where n corresponds to the order of A. Then n traces are obtained:

$$\mu_1 = \text{tr } A$$

$$\mu_2 = \text{tr } A^2$$

$$\mu_3 = \text{tr } A^3$$

$$\cdots$$

$$\mu_n = \text{tr } A^n$$

Then the coefficients of the characteristic polynomial $p(\lambda)$:

$$p(\lambda) = \lambda^n + \beta_1\lambda^{n-1} + \beta_2\lambda^{n-2} + \cdots + \beta_{n-2}\lambda^2 + \beta_{n-1}\lambda + \beta_n$$

are obtained from the sequence of operations:

$$\beta_1 = -\mu_1$$

$$\beta_2 = -\frac{(\mu_1\beta_1 + \mu_2)}{2}$$

$$\beta_3 = -\frac{(\mu_1\beta_2 + \mu_2\beta_1 + \mu_3)}{3}$$

$$\cdots$$

$$\beta_n = -\frac{(\mu_1\beta_{n-1} + \mu_2\beta_{n-2} + \cdots + \mu_{n-1}\beta_1 + \mu_n)}{n}$$

Example: With

$$A = \begin{bmatrix} 0 & 1 & 0 \\ 0 & 0 & 1 \\ -6 & -11 & -6 \end{bmatrix}$$

A^2 is obtained:

$$A^2 = \begin{bmatrix} 0 & 1 & 0 \\ 0 & 0 & 1 \\ -6 & -11 & -6 \end{bmatrix}\begin{bmatrix} 0 & 1 & 0 \\ 0 & 0 & 1 \\ -6 & -11 & -6 \end{bmatrix} = \begin{bmatrix} 0 & 0 & 1 \\ -6 & -11 & -6 \\ 36 & 60 & 25 \end{bmatrix}$$

and so is A^3:

$$A^3 = \begin{bmatrix} 0 & 1 & 0 \\ 0 & 0 & 1 \\ -6 & -11 & -6 \end{bmatrix}\begin{bmatrix} 0 & 0 & 1 \\ -6 & -11 & -6 \\ 36 & 60 & 25 \end{bmatrix} = \begin{bmatrix} -6 & -11 & -6 \\ 36 & 60 & 25 \\ -150 & -239 & -90 \end{bmatrix}$$

Then

$$\mu_1 = \text{tr } A = -6$$

$$\mu_2 = \text{tr } A = 14$$

$$\mu_3 = \text{tr } A = -36$$

and

$$\beta_1 = -\mu_1 = 6$$

$$\beta_2 = -\frac{1}{2}(\mu_1\beta_1 + \mu_2) = -\frac{1}{2}[-6(6) + 14] = -\frac{1}{2}(-22) = 11$$

and

$$\beta_3 = -\frac{1}{3}(\mu_1\beta_2 + \mu_2\beta_1 + \mu_3) = -\frac{1}{3}[-6(11) + 14(6) + (-36)] = 6$$

These values of β are the coefficients of $p(\lambda)$:

$$p(\lambda) = \lambda^3 + 6\lambda^2 + 11\lambda + 6$$

7.7 LEVERRIER'S METHOD†

Leverrier's method for the generation of the characteristic polynomial $p(\lambda)$ of the $n \times n$ matrix A also follows a prescribed procedure. The n coefficients of the polynomial

$$p(\lambda) = \lambda^n + \beta_1\lambda^{n-1} + \beta_2\lambda^{n-2} + \cdots + \beta_{n-2}\lambda^2 + \beta_{n-1}\lambda + \beta_n$$

are calculated in a rather simple procedure. The computation begins by setting $A_1 = A$, obtaining the trace of A, and then finding a coefficient:

$$\beta_1 = -\text{tr } A$$

Two new matrices are then formed:

$$B_1 = A_1 + \beta_1 I$$

and then

$$A_2 = AB_1$$

The procedure is then repeated n times, and the recurrence relationships are

$$\beta_i = -\frac{1}{i}\text{tr } A_i \quad (i = 1, 2, 3, \ldots, n)$$

$$B_i = A_i + \beta_i I \quad (i = 1, 2, 3, \ldots, n)$$

†Fadeev, D. K., and N. K. Fadeeva, *Computational Methods in Linear Algebra*, translated by R. C. Williams, W. H. Freeman, San Francisco, 1963.

and

$$A_k = AB_{k-1} \qquad (k = 1, 2, 3, \ldots, n)$$

Example: Again with A:

$$A = \begin{bmatrix} 0 & 1 & 0 \\ 0 & 0 & 1 \\ -6 & -11 & -6 \end{bmatrix}$$

tr $A = -6$ and $\beta_1 = -\text{tr}\, A = 6$. Then with $A_1 = A$

$$B_1 = A_1 + \beta_1 I = \begin{bmatrix} 0 & 1 & 0 \\ 0 & 0 & 1 \\ -6 & -11 & -6 \end{bmatrix} + \begin{bmatrix} 6 & 0 & 0 \\ 0 & 6 & 0 \\ 0 & 0 & 6 \end{bmatrix} = \begin{bmatrix} 6 & 1 & 0 \\ 0 & 6 & 1 \\ -6 & -11 & 0 \end{bmatrix}$$

and

$$A_2 = AB_1 = \begin{bmatrix} 0 & 1 & 0 \\ 0 & 0 & 1 \\ -6 & -11 & -6 \end{bmatrix}\begin{bmatrix} 6 & 1 & 0 \\ 0 & 6 & 1 \\ -6 & -11 & 0 \end{bmatrix}$$

$$= \begin{bmatrix} 0 & 6 & 1 \\ -6 & -11 & 0 \\ 0 & -6 & -11 \end{bmatrix}$$

Next,

$$\beta_2 = -\frac{1}{2}\, \text{tr}\, A_2 = -\frac{1}{2}(-22) = 11$$

$$B_2 = A_2 + \beta_2 I = \begin{bmatrix} 0 & 6 & 1 \\ -6 & -11 & 0 \\ 0 & -6 & -11 \end{bmatrix} + \begin{bmatrix} 11 & 0 & 0 \\ 0 & 11 & 0 \\ 0 & 0 & 11 \end{bmatrix}$$

$$= \begin{bmatrix} 11 & 6 & 1 \\ -6 & 0 & 0 \\ 0 & -6 & 0 \end{bmatrix}$$

and

$$A_3 = AB_2 = \begin{bmatrix} 0 & 1 & 0 \\ 0 & 0 & 1 \\ -6 & -11 & -6 \end{bmatrix}\begin{bmatrix} 11 & 6 & 1 \\ -6 & 0 & 0 \\ 0 & -6 & 0 \end{bmatrix} = \begin{bmatrix} -6 & 0 & 0 \\ 0 & -6 & 0 \\ 0 & 0 & -6 \end{bmatrix}$$

The final step is the determination of

$$\beta_3 = -\frac{1}{3}\,\text{tr}\,A_3 = -\frac{1}{3}(-18) = 6$$

and the characteristic polynomial is

$$p(\lambda) = \lambda^3 + 6\lambda^2 + 11\lambda + 6$$

Leverrier's method also provides the inverse of A. This bonus comes from a simple calculation involving B_{n-1} and β_n:

$$A^{-1} = -\frac{1}{\beta_n}B_{n-1}$$

Example: With A, B_2, and β_3 taken from the previous example,

$$A^{-1} = -\frac{1}{\beta_3}B_2 = -\frac{1}{6}\begin{bmatrix} 11 & 6 & 1 \\ -6 & 0 & 0 \\ 0 & -6 & 0 \end{bmatrix}$$

and it is easy to see that

$$AA^{-1} = \begin{bmatrix} 0 & 1 & 0 \\ 0 & 0 & 1 \\ -6 & -11 & -6 \end{bmatrix}\begin{bmatrix} -11/6 & -1 & -1/6 \\ 1 & 0 & 0 \\ 0 & 1 & 0 \end{bmatrix} = \begin{bmatrix} 1 & 0 & 0 \\ 0 & 1 & 0 \\ 0 & 0 & 1 \end{bmatrix}$$

7.8 EIGENVECTORS FOR NONREPEATED EIGENVALUES

Every eigenvalue leads to a solution of Eq. (7.1). If the eigenvalues are nonrepeated, then

$$[A - \lambda_k I]X_k = 0 \qquad (k = 1, 2, 3, \ldots, n) \tag{7.9}$$

where A is $n \times n$ and where the λ_ks are the n characteristic or eigenvalues of A. Because each X is $n \times 1$, Eq. (7.9) represents a set of n linear, simultaneous homogeneous algebraic equations (see Section 5.6), and because the rank of $[A - \lambda_k I]$ is equal to the rank of the augmented matrix $[A - \lambda_k I{:}0]$, a nontrivial solution exists if and only if det $[A - \lambda_k I] = 0$.

The solution vector X_k for each eigenvalue is called a *characteristic vector* or *eigenvector* and the set of n eigenvectors that are all $n \times 1$ constitute a linearly independent set if all the eigenvalues are distinct or unequal. This is easily proved by mathematical induction.

Suppose, for example, that there are two eigenvectors that derive from two unequal eigenvalues and that some linear combination of them yields a zero:

$$\alpha_1 X_1 + \alpha_2 X_2 = 0$$

If this is premultiplied by A, then

$$\alpha_1 A X_1 + \alpha_2 A X_2 = 0$$

and by noting that $AX = \lambda X$, this may be written as

$$\alpha_1 \lambda_1 X_1 + \alpha_2 \lambda_2 X_2 = 0$$

If the first equation is multiplied by λ_2, a subtraction may then be effected:

$$(\alpha_1 \lambda_1 X_1 + \alpha_2 \lambda_2 X_2) - (\alpha_1 \lambda_2 X_1 + \alpha_2 \lambda_2 X_2) = \alpha_1(\lambda_1 - \lambda_2)X_1 = 0$$

But it has been postulated in the beginning that no X_k is null and that all eigenvalues are unequal. Thus $\lambda_1 - \lambda_2 \neq 0$ and $X_k \neq 0$. The only alternative is that $\alpha_1 = 0$, and an identical procedure can be used to show that $\alpha_2 = 0$. Moreover, a similar procedure (admittedly with more algebra) can extend this procedure to any number of the eigenvectors in the set. This reasoning, known as mathematical induction, shows that all eigenvectors that derive from unequal eigenvalues are linearly independent.

The actual determination of the eigenvectors is accomplished by placing the eigenvalue associated with the eigenvector into Eq. (7.9) and solving the resulting system of equations.

Example: It is known that the matrix A:

$$A = \begin{bmatrix} 0 & 1 & 0 \\ 0 & 0 & 1 \\ -6 & -11 & -6 \end{bmatrix}$$

possesses three eigenvalues that may be listed in order with the largest value first (see Section 7.2):

$$\lambda_1 = -1$$

$$\lambda_2 = -2$$

and

$$\lambda_3 = -3$$

The matrix $[A - \lambda I]$ is

$$[A - \lambda I] = \begin{bmatrix} -\lambda & 1 & 0 \\ 0 & -\lambda & 1 \\ -6 & -11 & -(6 + \lambda) \end{bmatrix}$$

and the eigenvector X_1 can be determined by using $\lambda_1 = -1$ in this matrix and forming a system of three homogeneous, linear simultaneous equations:

$$[A - \lambda I]X = \begin{bmatrix} 1 & 1 & 0 \\ 0 & 1 & 1 \\ -6 & -11 & -5 \end{bmatrix} \begin{bmatrix} x_1 \\ x_2 \\ x_3 \end{bmatrix} = \begin{bmatrix} 0 \\ 0 \\ 0 \end{bmatrix}$$

and this can be expanded to yield the linearly dependent system:

$$x_1 + x_2 \qquad = 0$$
$$x_2 + x_3 = 0$$
$$-6x_1 - 11x_2 - 5x_3 = 0$$

the solution here is

$$x_1 = -x_2$$

and

$$x_3 = -x_2$$

Thus with an arbitrary selection of $x_2 = 1$, the first eigenvector is

$$X_1 = \alpha_1 \begin{bmatrix} -1 \\ 1 \\ -1 \end{bmatrix}$$

For $\lambda = -2$,

$$[A - \lambda I]X = \begin{bmatrix} 2 & 1 & 0 \\ 0 & 2 & 1 \\ -6 & -11 & -4 \end{bmatrix} \begin{bmatrix} x_1 \\ x_2 \\ x_3 \end{bmatrix} = \begin{bmatrix} 0 \\ 0 \\ 0 \end{bmatrix}$$

or

$$2x_1 + x_2 \qquad = 0$$
$$2x_2 + x_3 = 0$$
$$-6x_1 - 11x_2 - 4x_3 = 0$$

which is another linearly dependent system. Here

$$x_1 = -\frac{x_2}{2}$$

and

$$x_3 = -2 x_2$$

so that with the arbitrary selection of $x_2 = 1$, the second eigenvector becomes:

$$X_2 = \alpha_2 \begin{bmatrix} -1/2 \\ 1 \\ -2 \end{bmatrix}$$

Finally, with $\lambda_3 = -3$

$$[A - I]X = \begin{bmatrix} 3 & 1 & 0 \\ 0 & 3 & 1 \\ -6 & -11 & -3 \end{bmatrix} \begin{bmatrix} x_1 \\ x_2 \\ x_3 \end{bmatrix} = \begin{bmatrix} 0 \\ 0 \\ 0 \end{bmatrix}$$

or in expanded form,

$$3x_1 + x_2 = 0$$
$$3x_2 + x_3 = 0$$
$$-6x_1 - 11x_2 - 3x_3 = 0$$

Here

$$x_1 = -\frac{x_2}{3}$$

and

$$x_3 = -3x_2$$

so that with the arbitrary selection of $x_2 = 1$, the third eigenvector becomes:

$$X_3 = \alpha_3 \begin{bmatrix} -1/3 \\ 1 \\ -3 \end{bmatrix}$$

The reader may verify that the three eigenvectors do indeed form a linearly independent set.

7.9 EIGENVECTORS FOR REPEATED EIGENVALUES

When eigenvalues of the matrix A are repeated with a multiplicity of r, some of the eigenvalues may be linearly dependent on others. Guidance on the number of linearly independent eigenvectors can be obtained from the rank of the matrix A.

As shown in Sections 5.6 and 5.8, a set of simultaneous, linear, homogeneous algebraic equations, if consistent, produces a unique solution if the rank of the $n \times n$ coefficient matrix is equal to its order. If the rank of the coefficient matrix is less than its order, an infinite number of solutions is produced.

To determine how many linearly independent eigenvectors are associated with each repeated eigenvalue, it is necessary to examine the rank of the matrix $K = [A - \lambda I]$. The first step is to form K with the repeated eigenvalue inserted. Then the rank of K is determined and the number of linearly independent eigenvectors associated with the repeated eigenvalue is equal to the difference between the order of K and the rank of A; that is, $n - r$.

Example: For the matrix

$$A = \begin{bmatrix} 0 & 2 & 0 \\ 2 & 0 & 0 \\ 0 & 0 & 2 \end{bmatrix}$$

the characteristic matrix is

$$K = [A - \lambda I] = \begin{bmatrix} -\lambda & 2 & 0 \\ 2 & -\lambda & 0 \\ 0 & 0 & (2 - \lambda) \end{bmatrix}$$

The characteristic equation is obtained by setting the determinant of the characteristic matrix equal to zero:

$$p(\lambda) = \det [A - \lambda I] = \lambda^2(2 - \lambda) - 4(2 - \lambda) = (2 - \lambda)(\lambda^2 - 4) = 0$$

and this yields three eigenvalues, one of which is repeated:

$$\lambda_1 = 2$$

$$\lambda_2 = 2$$

and

$$\lambda_3 = -2$$

With $\lambda_1 = \lambda_2 = 2$, $[A - \lambda I]X = 0$ becomes

$$[A - \lambda I]X = \begin{bmatrix} -2 & 2 & 0 \\ 2 & -2 & 0 \\ 0 & 0 & 0 \end{bmatrix} \begin{bmatrix} x_1 \\ x_2 \\ x_3 \end{bmatrix} = \begin{bmatrix} 0 \\ 0 \\ 0 \end{bmatrix}$$

and for this particular K which is 3×3 ($n = 3$), it is easy to see that the rank of K is $r(K) = 1$. Thus with $n - r = 3 - 1 = 2$, two linearly independent eigenvectors are obtained. Expansion of the foregoing matrix representation yields two linearly dependent equations:

$$-2x_1 + 2x_2 = 0$$

$$2x_1 - 2x_2 = 0$$

which, from either, provides $x_1 = x_2$. This means that x_3 can be arbitrary (which shows that an infinite number of solutions exists).

The two linearly independent eigenvectors can be obtained by setting x_1 (say $x_1 = 1$) and letting $x_3 = 0$ or by arbitrarily setting x_3 (say $x_3 = 1$) and calling $x_1 = x_2 = 0$. In this event, the two eigenvectors are:

$$X_1 = \alpha_1 \begin{bmatrix} 1 \\ 1 \\ 0 \end{bmatrix} \qquad X_2 = \alpha_2 \begin{bmatrix} 0 \\ 0 \\ 1 \end{bmatrix}$$

and it is easy to see that these are linearly independent.

The third eigenvalue is $\lambda_3 = -2$. With this in the characteristic matrix $K = [A - \lambda I]$, one obtains:

$$[A - \lambda I] = \begin{bmatrix} 2 & 2 & 0 \\ 2 & 2 & 0 \\ 0 & 0 & 4 \end{bmatrix}$$

and the set of simultaneous, linear homogeneous equations can be represented by

$$[A - \lambda I]X = \begin{bmatrix} 2 & 2 & 0 \\ 2 & 2 & 0 \\ 0 & 0 & 2 \end{bmatrix} \begin{bmatrix} x_1 \\ x_2 \\ x_3 \end{bmatrix} = \begin{bmatrix} 0 \\ 0 \\ 0 \end{bmatrix}$$

or in expanded form as

$$2x_1 + 2x_2 = 0$$

$$2x_1 + 2x_2 = 0$$

$$4x_3 = 0$$

This time, $x_3 = 0$ and $x_1 = -x_2$ so that another linearly independent vector comes forth:

$$X_3 = \alpha_3 \begin{bmatrix} 1 \\ -1 \\ 0 \end{bmatrix}$$

In summary, the three linearly independent eigenvectors are:

$$X_1 = \alpha_1 \begin{bmatrix} 1 \\ 1 \\ 0 \end{bmatrix} \qquad X_2 = \alpha_2 \begin{bmatrix} 0 \\ 0 \\ 1 \end{bmatrix} \qquad X_3 = \alpha_3 \begin{bmatrix} 1 \\ -1 \\ 0 \end{bmatrix}$$

Example: It is easy to see that the matrix A:

$$A = \begin{bmatrix} 4 & 0 & 0 \\ 0 & 4 & 0 \\ 0 & 0 & 4 \end{bmatrix}$$

has a characteristic matrix:

$$K = [A - \lambda I] = \begin{bmatrix} (4 - \lambda) & 0 & 0 \\ 0 & (4 - \lambda) & 0 \\ 0 & 0 & (4 - \lambda) \end{bmatrix}$$

a characteristic polynomial:

$$p(\lambda) = (4 - \lambda)^3 = 0$$

and three identical eigenvalues, $\lambda_1 = \lambda_2 = \lambda_3 = 4$. When these eigenvalues are put into the characteristic matrix, a square null matrix of order $n = 3$ and rank $r(K) = 0$ results. With $3 - 0 = 3$, one expects that there will be three linearly independent eigenvectors.

Observe that when the eigenvectors are inserted into K, the result is a set of homogeneous equations:

$$[A - \lambda I]X = \begin{bmatrix} 0 & 0 & 0 \\ 0 & 0 & 0 \\ 0 & 0 & 0 \end{bmatrix} \begin{bmatrix} x_1 \\ x_2 \\ x_3 \end{bmatrix} = \begin{bmatrix} 0 \\ 0 \\ 0 \end{bmatrix}$$

and after these are expanded, one obtains

$$(0)x_1 = 0$$

$$(0)x_2 = 0$$

and

$$(0)x_3 = 0$$

Thus the eigenvectors X_1, X_2, and X_3 are all perfectly arbitrary and linearly independent. Any linearly independent set would be a set of eigenvectors, and here is just one of them:

$$X_1 = \begin{bmatrix} 1 \\ 0 \\ 0 \end{bmatrix} \quad X_2 = \begin{bmatrix} 0 \\ 1 \\ 0 \end{bmatrix} \quad X_3 = \begin{bmatrix} 0 \\ 0 \\ 1 \end{bmatrix}$$

Example: As a final example, consider

$$A = \begin{bmatrix} 4 & -2 \\ 0 & 4 \end{bmatrix}$$

that has a characteristic matrix:

$$K = [A - \lambda I] = \begin{bmatrix} (4 - \lambda) & -2 \\ 0 & (4 - \lambda) \end{bmatrix}$$

The characteristic equation derives from $\det K = 0$:

$$\det K = (4 - \lambda)^2 = 0$$

and this shows that the two eigenvalues are identical:

$$\lambda_1 = 4$$

and

$$\lambda_2 = 4$$

When these eigenvalues are put into K, the resulting system of equations is given by

$$[A - I]X = \begin{bmatrix} 0 & -2 \\ 0 & 0 \end{bmatrix} \begin{bmatrix} x_1 \\ x_2 \end{bmatrix} = \begin{bmatrix} 0 \\ 0 \end{bmatrix}$$

and from that it is observed that $[A - \lambda I]$ is 2×2 ($n = 2$) but $r([A - \lambda I]) = 1$. Thus $n - r = 2 - 1 = 1$, and one linearly independent eigenvector is to be expected.

An expansion of $[A - \lambda I]X$ gives one equation:

$$(0)x_1 - 2x_2 = 0$$

where x_1 may be arbitrary and $x_2 = 0$. Thus one eigenvector is:

$$X_1 = \alpha_1 \begin{bmatrix} 1 \\ 0 \end{bmatrix}$$

and any other eigenvector may be formed by using a different scaling factor:

$$X_2 = \alpha_2 \begin{bmatrix} 1 \\ 0 \end{bmatrix}$$

The vectors so obtained, because they differ by a constant, are proportional to one another and are therefore linearly dependent.

7.10 THE MODAL MATRIX

A square matrix of order n can be formed from the eigenvectors of the $n \times n$ square matrix A. This vector is called the *modal matrix* of A and is designated by M:

$$M = [X_1\ X_2\ X_3 \cdots X_n] \qquad (7.10)$$

Example: The matrix

$$A = \begin{bmatrix} 0 & 1 & 6 \\ 0 & 0 & 1 \\ -6 & -11 & -6 \end{bmatrix}$$

has been shown in the example in Sec. 7.8 to possess three nonrepeated eigenvalues $\lambda_1 = -1$, $\lambda_2 = -2$, and $\lambda_3 = -3$. Each of these eigenvalues is associated with an eigenvector:

$$X_1 = \alpha_1 \begin{bmatrix} -1 \\ 1 \\ -1 \end{bmatrix} \quad X_2 = \alpha_2 \begin{bmatrix} -1/2 \\ 1 \\ -2 \end{bmatrix} \quad \text{and} \quad X_3 = \alpha_3 \begin{bmatrix} -1/3 \\ 1 \\ -3 \end{bmatrix}$$

This linearly independent set of column vectors can be used to form the modal matrix of A (setting $\alpha_1 = \alpha_2 = \alpha_3 = 1$)

$$M = \begin{bmatrix} -1 & -1/2 & -1/3 \\ 1 & 1 & 1 \\ -1 & -2 & -3 \end{bmatrix}$$

7.11 EIGENVALUES AND EIGENVECTORS FOR SYMMETRICAL MATRICES

In the event that the $n \times n$ square matrix A is symmetric such that $A^T = A$, there are certain other properties in addition to those already considered.

The unsymmetric matrix may possess real or complex eigenvalues. However, the eigenvalues of symmetric matrices, which are the inevitable occurrence in many types of engineering analyses in a wide range of applications, are always real. Moreover, the symmetrix matrix always possesses linearly independent eigenvectors and this is guaranteed in all cases in which the eigenvalues are repeated.

To show that the eigenvalues of a symmetric matrix are real, let H be an eigenvector of the $n \times n$ matrix A that possesses the property that $A^T = A$ (A is symmetric). Then, in accordance with Eq. (7.1),

$$AH = \lambda H \qquad (7.11a)$$

where λ is the eigenvalue that leads to H. If λ is a complex number but the elements of A are real, it is easy to see that at least some (if not all) the elements of H are complex numbers. If all the complex elements of Eq. (7.11a) are replaced by their conjugates (remember, A is real and at this point λ may be complex)

$$AH^* = \lambda^* H^* \qquad (7.11b)$$

If Eq. (7.11a) is multiplied by H^{*T} and Eq. (7.11b) is multiplied by H^T, the result is a pair of equations:

$$H^{*T}AH = \lambda H^{*T}H \qquad (7.12a)$$

and

$$H^TAH^* = \lambda^* H^TH^* \qquad (7.12b)$$

The transpose of Eq. (7.12b) may then be taken:

$$H^{*T}A^TH = \lambda^* H^{*T}H \qquad (7.12c)$$

and with the recognition that $A^T = A$, a subtraction of Eq. (7.12a) from Eq. (7.12c) provides

$$\lambda^* H^{*T}H = \lambda H^{*T}H$$

which, because $H^{*T}H \neq 0$, indicates that

$$\lambda^* = \lambda$$

which is a fact only if λ is real.

The eigenvectors associated with unequal eigenvalues are orthogonal. Consider λ_j and λ_k (with associated eigenvectors X_j and X_k) as two unequal eigenvalues so that, in accordance with Eq. (7.1)

$$AX_j = \lambda_j X_j$$

$$AX_k = \lambda_k X_k$$

If the first of these is multiplied by X_k^T and the second by X_j^T (these transposes are sometimes referred to as *eigenrows*) the result is:

$$X_k^T A X_j = \lambda_j X_k^T X_j$$

$$X_j^T A X_k = \lambda_k X_j^T X_k$$

and the transposes of these yield $(A^T = A)$

$$X_j^T A X_k = \lambda_j X_j^T X_k$$

$$X_k^T A X_j = \lambda_k X_k^T X_j$$

These four equations provide two equalities, from the first and the fourth:

$$\lambda_j X_k^T X_j = \lambda_k X_k^T X_j$$

and from the second and the third:

$$\lambda_k X_j^T X_k = \lambda_j X_j^T X_k$$

Both of these may be adjusted algebraically to

$$(\lambda_j - \lambda_k) X_k^T X_j = 0$$

and

$$(\lambda_k - \lambda_j) X_j^T X_k = 0$$

But it was presumed originally that the two eigenvalues λ_1 and λ_2 were unequal. This means that

$$X_k^T X_j = X_j^T X_k = 0$$

and proves that the two eigenvectors X_j and X_k formed from the unequal eigenvalues are orthogonal. Moreover, this means that the modal matrix constructed from the eigenvalues of unequal eigenvectors of a symmetric matrix is orthogonal.

Example: The matrix A:

$$A = \begin{bmatrix} 4 & -2 & 0 \\ -2 & 3 & -1 \\ 0 & -1 & 2 \end{bmatrix}$$

is symmetric and has a characteristic matrix:

$$K = [A - \lambda I] = \begin{bmatrix} (4 - \lambda) & -2 & 0 \\ -2 & (3 - \lambda) & -1 \\ 0 & -1 & (2 - \lambda) \end{bmatrix}$$

and a characteristic equation derived from det $K = 0$:

$$p(\lambda) = (4 - \lambda)(3 - \lambda)(2 - \lambda) - (4 - \lambda) - 4(2 - \lambda) = 0$$
$$= -\lambda^3 + 9\lambda^2 - 26\lambda + 24 - 12 + 5\lambda = 0$$

or

$$p(\lambda) = \lambda^3 - 9\lambda^2 + 21\lambda - 12 = 0$$

This equation has three roots that are the eigenvalues of A:

$$\lambda_1 = 5.6691$$

$$\lambda_2 = 2.4760$$

and

$$\lambda_3 = 0.8549$$

and these are observed to be real, as predicted, and unequal.
   The eigenvector associated with $\lambda_1 = 5.6691$ derives from

$$[A - \lambda I]X - \begin{bmatrix} -1.6691 & -2 & 0 \\ -2 & -2.6691 & -1 \\ 0 & -1 & -3.6691 \end{bmatrix} \begin{bmatrix} x_1 \\ x_2 \\ x_3 \end{bmatrix} = \begin{bmatrix} 0 \\ 0 \\ 0 \end{bmatrix}$$

which expands to the homogeneous system

$$-1.6691x_1 - 2x_2 = 0$$
$$-2x_1 - 2.6691x_2 - x_3 = 0$$
$$-x_2 - 3.6691x_3 = 0$$

and says that

$$x_1 = -1.1983x_2$$

and

$$x_3 = -0.2725x_2$$

Thus by selecting the arbitrary value $x_2 = 1.0000$, the first eigenvector corresponding to the eigenvalue $\lambda_1 = 5.6691$ becomes

$$X_1 = \alpha_1 \begin{bmatrix} -1.1983 \\ 1.0000 \\ -0.2725 \end{bmatrix}$$

The eigenvector associated with $\lambda_2 = 2.4760$ comes from the matrix representation:

$$[A - \lambda I]X = \begin{bmatrix} 1.5240 & -2 & 0 \\ -2 & 0.5240 & -1 \\ 0 & -1 & -0.4760 \end{bmatrix} \begin{bmatrix} x_1 \\ x_2 \\ x_3 \end{bmatrix} = \begin{bmatrix} 0 \\ 0 \\ 0 \end{bmatrix}$$

or in expanded form:

$$1.5420x_1 \quad - 2x_2 \qquad\qquad = 0$$
$$-2x_1 + 0.5240x_2 \qquad - x_3 = 0$$
$$- x_2 - 0.4760x_3 = 0$$

from which

$$x_1 = 1.3124x_2$$

and

$$x_3 = -2.1007x_2$$

This time, the arbitrary value of $x_2 = 1.0000$ yields the eigenvector corresponding to $\lambda_2 = 2.4760$:

$$X_2 = \alpha_2 \begin{bmatrix} 1.3124 \\ 1.0000 \\ -2.1007 \end{bmatrix}$$

Finally, with $\lambda_3 = 0.8549$

$$[A - \lambda I]X = \begin{bmatrix} 3.1451 & -2 & 0 \\ -2 & 2.1451 & -1 \\ 0 & -1 & 1.1451 \end{bmatrix} \begin{bmatrix} x_1 \\ x_2 \\ x_3 \end{bmatrix} = \begin{bmatrix} 0 \\ 0 \\ 0 \end{bmatrix}$$

This expands to

$$3.1451x_1 \quad - 2x_2 \qquad\qquad = 0$$
$$-2x_1 + 2.1451x_2 \qquad - x_3 = 0$$
$$-x_2 + 1.1451x_3 = 0$$

and yields

$$x_1 = 0.6359x_2$$

and

$$x_3 = 0.8733x_2$$

Again with $x_2 = 1.0000$ (arbitrary), the eigenvector corresponding to $\lambda_3 = 0.8549$ is:

$$X_3 = \alpha_3 \begin{bmatrix} 0.6359 \\ 1.0000 \\ 0.8733 \end{bmatrix}$$

The alphas in the three eigenvectors are arbitrary constants. When they are set to unity, the modal matrix may be written as:

$$M = \begin{bmatrix} -1.1983 & 1.3124 & 0.6359 \\ 1.0000 & 1.0000 & 1.0000 \\ -0.2725 & -2.1007 & 0.8733 \end{bmatrix}$$

and the reader may verify that the three 3×1 vectors in M are both linearly independent and orthogonal.

7.12 ADDITIONAL PROPERTIES OF EIGENVALUES AND EIGENVECTORS

Equation (7.1) may be multiplied by a constant, say β. Thus

$$\beta(AX) = \beta(\lambda X)$$

which may be written as

$$(\beta A)X = BX = (\beta\lambda)X = \mu X$$

The $n \times n$ matrix A, whether symmetrical or not, has been multiplied by a scalar, as has its eigenvalues. The result is a new matrix B, with a different set of eigenvalues μ, which are related to the eigenvalues of A by the same constant or scalar that transformed A into B. Thus it can be stated that if a matrix is multiplied by a scalar, its eigenvalues are multiplied by the same scalar.

Now take Eq. (7.1) and presume that A is nonsingular so that both sides may be premultiplied by the inverse of A:

$$A^{-1}AX = \lambda A^{-1}X$$

or

$$A^{-1}X = \frac{1}{\lambda}X$$

It is easy to see that this has the form of Eq. (7.1) but that A has been replaced by A^{-1} and λ has been replaced by $1/\lambda$. This indicates that the eigenvalues of the inverse of a nonsingular matrix A are the reciprocals of the eigenvalues of A and the eigenvectors of A and A^{-1} are the same; that is, the eigenvector of A that derives from λ is identical to the eigenvector of A^{-1} that derives from $1/\lambda$.

Once again, consider Eq. (7.1):

$$AX = \lambda X$$

and compare this to a version where A^T is employed

$$A^T Y = \mu Y$$

Here, the eigenvalues of A^T that are designated by μ are obtained from

$$\det [A^T - \mu I] = 0$$

However, $\det A = \det A^T$ and $I = I^T$ so that

$$\det [A^T - \mu I^T] = \det [A - \lambda I] = 0$$

which produces the same characteristic equation and which shows that the μ's deriving from A^T are equal to the λ's deriving from A. Thus the eigenvalues of A are equal to the eigenvalues of A^T, but the eigenvectors may be different.

Example: Once again consider the matrix

$$A = \begin{bmatrix} 0 & 1 & 0 \\ 0 & 0 & 1 \\ -6 & -11 & -6 \end{bmatrix}$$

which is known to have a characteristic equation (Section 7.2):

$$p(\lambda) = \lambda^3 + 6\lambda + 11\lambda + 6 = 0$$

three eigenvalues $\lambda_1 = -1$, $\lambda_2 = -2$, and $\lambda_3 = -3$; and three eigenvectors (Section 7.8):

$$X_1 = \alpha_1 \begin{bmatrix} -1 \\ 1 \\ -1 \end{bmatrix}, X_2 = \alpha_2 \begin{bmatrix} -1/2 \\ 1 \\ -2 \end{bmatrix} \text{ and } X_3 = \alpha_3 \begin{bmatrix} -1/3 \\ 1 \\ -3 \end{bmatrix}$$

The transpose of A is:

$$A^T = \begin{bmatrix} 0 & 0 & -6 \\ 1 & 0 & -11 \\ 0 & 1 & -6 \end{bmatrix}$$

and A^T has a characteristic matrix:

$$K = [A^T - \lambda I] = \begin{bmatrix} -\lambda & 0 & -6 \\ 1 & -\lambda & -11 \\ 0 & 1 & -(6 + \lambda) \end{bmatrix}$$

With det $K = \det [A^T - \lambda I] = 0$,

$$p(\lambda) = -\lambda^2(6 + \lambda) - 6 - 11\lambda = 0$$
$$= -\lambda^3 - 6\lambda^2 - 11\lambda - 6 = 0$$

or

$$p(\lambda) = \lambda^3 + 6\lambda^2 + 11\lambda + 6 = 0$$

which is the same characteristic equation as obtained from the evaluation of det $[A - \lambda I] = 0$. Thus the eigenvalues of A^T are the same as those for A, and it remains to see whether the eigenvectors of A^T differ from those of A.

For the first eigenvalue $\lambda_1 = -1$,

$$[A^T - \lambda I]X = \begin{bmatrix} 1 & 0 & -6 \\ 1 & 1 & -11 \\ 0 & 1 & -5 \end{bmatrix} \begin{bmatrix} x_2 \\ x_2 \\ x_3 \end{bmatrix} = \begin{bmatrix} 0 \\ 0 \\ 0 \end{bmatrix}$$

which can be expanded to yield:

$$x_1 \quad - \quad 6x_3 = 0$$
$$x_1 + x_2 - 11x_3 = 0$$
$$x_2 - \quad 5x_3 = 0$$

from which

$$x_1 = 6x_3$$

and

$$x_2 = 5x_3$$

If x_3 is selected arbitrarily as $x_3 = 1$, then the eigenvector corresponding to $\lambda = -1$ is:

$$X_1 = \alpha_1 \begin{bmatrix} 6 \\ 5 \\ 1 \end{bmatrix}$$

The reader may wish to verify that for $\lambda_2 = -2$ and $\lambda_2 = -3$, the eigenvectors are:

$$X_2 = \alpha_2 \begin{bmatrix} 3 \\ 4 \\ 1 \end{bmatrix}; \qquad X_3 = \alpha_3 \begin{bmatrix} 2 \\ 3 \\ 1 \end{bmatrix}$$

and it is seen that none of the eigenvectors of A^T bear any resemblance to those of A.

If the $n \times n$ matrix A is orthogonal, then $A^T = A^{-1}$. By Eq. (7.1),

$$AX = \lambda X$$

and

$$(AX)^T = (\lambda X)^T$$

or

$$X^T A^T = \lambda X^T$$

The product $(AX)^T(AX)$ may be written by following either of two procedures:

$$(AX)^T(AX) = (X^T A^T)AX = X^T(A^T A)X = X^T I X = X^T X$$

and

$$(AX)^T(AX) = (\lambda X)^T(\lambda X) = \lambda^2 X^T X$$

A comparison of these shows that if $X^T X \neq 0$, $\lambda^2 = 1$, and $\lambda = \pm 1$. This means that all the eigenvalues of orthogonal matrices are either $+1$ or -1.

Diagonal and triangular matrices (whether upper triangular or lower triangular) possess the property that their determinants are equal to the product of the elements that are on their principal diagonal. Thus their eigenvalues are equal to the elements on the principal diagonal.

7.13 THE DIAGONAL FORM OF A MATRIX

Consider an $n \times n$ matrix A and its modal matrix M. M is composed of n linearly independent eigenvectors if A is symmetrical. If A is not symmetrical, the eigenvalues of A are unequal. If M is formed from the n linearly independent eigenvectors of A, then each eigenvector of M satisfies Eq. (7.1) which may be written for the kth eigenvalue and eigenvector as

$$AX_k = \lambda_k X_k$$

If A is postmultiplied by M, then

$$AM = A[X_1 \quad X_2 \quad X_3 \cdots X_n]$$

$$AM = [AX_1 \quad AX_2 \quad AX_3 \cdots AX_n]$$

or

$$AM = [\lambda_1 X_1 \quad \lambda_2 X_2 \quad \lambda_3 X_3 \cdots \lambda_n X_n] \tag{7.13}$$

Now define a *spectral matrix*, a diagonal matrix containing the eigenvalues of A on its principal diagonal:

$$S = \begin{bmatrix} \lambda_1 & 0 & 0 & \cdots & 0 \\ 0 & \lambda_2 & 0 & \cdots & 0 \\ 0 & 0 & \lambda_3 & \cdots & 0 \\ & & \cdots & & \\ 0 & 0 & 0 & \cdots & \lambda_n \end{bmatrix}$$

One may think of this spectral matrix as being derived from a sequence of elementary transformations. With this in mind, one may get the feel for the formation of Eq. (7.13) via a postmultiplication of the columns of M (the eigenvectors) by a sequence of elementary transformations represented by S. Hence

$$AM = MS \tag{7.14}$$

Because M is composed of linearly independent eigenvectors, M^{-1} exists and can be used to premultiply Eq. (7.14):

$$M^{-1}AM = S \tag{7.15}$$

This important relationship is a similarity transformation (see Section 6.10) and represents the transformation of A into a diagonal matrix. It is often referred to as the diagonalization of A.

There are some implications here that should not be overlooked. The first is the temptation of merely writing the spectral matrix using known eigenvalues without bothering to form the modal matrix. This, of course, can be done, but in doing so, one may miss the important engineering applications of the diagonalization process. The diagonalization process, however, is not a practical way of finding the spectral matrix, if the spectral matrix is all that is required, because the eigenvalues that compose the spectral matrix are needed to determine M.

Any $n \times n$ matrix that possesses n linearly independent eigenvectors is diagonalizable. The matrix A with unequal eigenvalues and the symmetric matrix A are guaranteed to be diagonalizable. It has been observed that not all $n \times n$ matrices possess linearly independent eigenvectors, and this is the criterion. Some matrices with repeated eigenvalues have linearly independent eigenvectors, some do not. The criterion does not pertain to repeated eigenvalues.

Finally, it must be recognized that because M is not unique (each eigenvector is multiplied by an arbitrary constant), if an eigenvector is multiplied by a constant, the new modal matrix M so formed still produces the diagonalization.

Example: The matrix A:

$$A = \begin{bmatrix} 0 & 1 & 0 \\ 0 & 0 & 1 \\ -6 & -11 & -6 \end{bmatrix}$$

is known to have a modal matrix composed of three eigenvectors (Section 7.10):

$$M = \begin{bmatrix} -1 & -1/2 & -1/3 \\ 1 & 1 & 1 \\ -1 & -2 & -3 \end{bmatrix}$$

and this modal matrix possesses an inverse, which the reader may verify:

$$M^{-1} = \begin{bmatrix} -3 & -5/2 & -1/2 \\ 6 & 8 & 2 \\ -3 & -9/2 & -3/2 \end{bmatrix}$$

The diagonalization of A now proceeds:

$$M^{-1}AM = \begin{bmatrix} -3 & -5/2 & -1/2 \\ 6 & 8 & 2 \\ -3 & -9/2 & -3/2 \end{bmatrix} \begin{bmatrix} 0 & 1 & 0 \\ 0 & 0 & 1 \\ -6 & -11 & -6 \end{bmatrix} \begin{bmatrix} -1 & -1/2 & -1/3 \\ 1 & 1 & 1 \\ -1 & -2 & -3 \end{bmatrix}$$

$$= \begin{bmatrix} 3 & 5/2 & 1/2 \\ -12 & -16 & -4 \\ 9 & 27/2 & 9/2 \end{bmatrix} \begin{bmatrix} -1 & -1/2 & -1/3 \\ 1 & 1 & 1 \\ -1 & -2 & -3 \end{bmatrix}$$

or

$$S = M^{-1}AM = \begin{bmatrix} -1 & 0 & 0 \\ 0 & -2 & 0 \\ 0 & 0 & -3 \end{bmatrix}$$

7.14 POWERS AND ROOTS OF MATRICES

The diagonalization of the $n \times n$ matrix A is the clue to an easy method for obtaining powers and roots of A. Consider Eq. (7.15) with a premultiplication by M and a postmultiplication by M^{-1}.

$$MM^{-1}AMM^{-1} = IAI = MSM^{-1}$$

or

$$A = MSM^{-1}$$

With this relation for A in terms of its modal and spectral matrices, it is easy to find powers of A. For example, for A^2,

$$A^2 = (MSM^{-1})(MSM^{-1}) = MS(M^{-1}M)SM^{-1} = MS^2M^{-1}$$

and for A, mathematical induction indicates that

$$A^m = MS^mM^{-1} \tag{7.16}$$

Example: For the matrix A:

$$A = \begin{bmatrix} 0 & 1 & 0 \\ 0 & 0 & 1 \\ -6 & -11 & -6 \end{bmatrix}$$

with the known modal matrix M and its inverse M^{-1} (see the example in Section 7.13), it is easy to see that for $A^3 = MS^3M^{-1}$, use of the spectral matrix containing the eigenvalues of A on its diagonal

$$S = \begin{bmatrix} -1 & 0 & 0 \\ 0 & -2 & 0 \\ 0 & 0 & -3 \end{bmatrix}$$

provides

$$A^3 = \begin{bmatrix} -1 & -1/2 & -1/3 \\ 1 & 1 & 1 \\ -1 & -2 & -3 \end{bmatrix} \begin{bmatrix} -1 & 0 & 0 \\ 0 & -8 & 0 \\ 0 & 0 & -27 \end{bmatrix} \begin{bmatrix} -3 & -5/2 & -1/2 \\ 6 & 8 & 2 \\ -3 & -9/2 & -3/2 \end{bmatrix}$$

$$= \begin{bmatrix} 1 & 4 & 9 \\ -1 & -8 & -27 \\ 1 & 16 & 81 \end{bmatrix} \begin{bmatrix} -3 & -5/2 & -1/2 \\ 6 & 8 & 2 \\ -3 & -9/2 & -3/2 \end{bmatrix}$$

or

$$A^3 = \begin{bmatrix} -6 & -11 & -6 \\ 36 & 60 & 25 \\ -150 & -239 & -90 \end{bmatrix}$$

This can be verified by a direct computation of A^3 which, at the same time, yields a comparison of the labor involved in the use of the two alternate methods:

$$A^3 = \begin{bmatrix} 0 & 1 & 0 \\ 0 & 0 & 1 \\ -6 & -11 & -6 \end{bmatrix} \begin{bmatrix} 0 & 1 & 0 \\ 0 & 0 & 1 \\ -6 & -11 & -6 \end{bmatrix} \begin{bmatrix} 0 & 1 & 0 \\ 0 & 0 & 1 \\ -6 & -11 & -6 \end{bmatrix}$$

$$= \begin{bmatrix} 0 & 0 & 1 \\ -6 & -11 & -6 \\ 36 & 60 & 25 \end{bmatrix} \begin{bmatrix} 0 & 1 & 0 \\ 0 & 0 & 1 \\ -6 & -11 & -6 \end{bmatrix}$$

or

$$A^3 = \begin{bmatrix} -6 & -11 & -6 \\ 36 & 60 & 25 \\ -150 & -239 & -90 \end{bmatrix}$$

Both methods yield identical values for A and the labor involved is about equal; however, the requirement for the computation of higher powers of A would probably swing the pendulum in favor of the use of Eq. (7.16), particularly if the eigenvalues and eigenvectors of A could be determined expeditiously.

Now look at Eq. (7.16) and replace m with $1/m$. This yields a relationship that will provide the mth root of A

$$A^{1/m} = MS^{1/m}M^{-1} \tag{7.17}$$

and this is easy to verify. For example, if $B = A^{1/3}$, then by Eq. (7.17), $B = MS^{-1/3}M^{-1}$ and

$$A = B^3 = (MS^{1/3}M^{-1})(MS^{1/3}M^{-1})(MS^{1/3}M^{-1})$$

$$= MS^{1/3}(M^{-1}M)S^{1/3}(M^{-1}M)S^{1/3}M^{-1}$$

or $A = MSM^{-1}$ which is Eq. (7.16).

Example: Consider the 2×2 matrix:

$$A = \begin{bmatrix} 2 & -1 \\ -2 & 3 \end{bmatrix}$$

which has two eigenvalues, $\lambda_1 = 4$ and $\lambda_2 = 1$ and two eigenvectors

$$X_1 = \alpha_1 \begin{bmatrix} 1 \\ -2 \end{bmatrix} \qquad X_2 = \alpha_2 \begin{bmatrix} 1 \\ 1 \end{bmatrix}$$

which are linearly independent. The modal matrix may be formed by letting $\alpha_1 = \alpha_2 = 1$:

$$M = \begin{bmatrix} 1 & 1 \\ -2 & 1 \end{bmatrix}$$

and this has an inverse

$$M = \begin{bmatrix} 1/3 & -1/3 \\ 2/3 & 1/3 \end{bmatrix}$$

The matrix A possesses four (not two) square roots. Observe that, in accordance with Eq. (7.17),

$$A^{1/2} = \begin{bmatrix} 1 & 1 \\ -2 & 1 \end{bmatrix} \begin{bmatrix} \pm 2 & 0 \\ 0 & \pm 1 \end{bmatrix} \begin{bmatrix} 1/3 & -1/3 \\ 2/3 & 1/3 \end{bmatrix}$$

and there are four combinations of plus and minus signs that lead to the four square roots. One of them is:

$$A^{1/2} = \begin{bmatrix} 1 & 1 \\ -2 & 1 \end{bmatrix} \begin{bmatrix} 2 & 0 \\ 0 & 1 \end{bmatrix} \begin{bmatrix} 1/3 & -1/3 \\ 2/3 & 1/3 \end{bmatrix}$$

$$= \begin{bmatrix} 2 & 1 \\ -4 & 1 \end{bmatrix} \begin{bmatrix} 1/3 & -1/3 \\ 2/3 & 1/3 \end{bmatrix}$$

or

$$A^{1/2} = \begin{bmatrix} 4/3 & -1/3 \\ -2/3 & 5/3 \end{bmatrix}$$

This is easily verified:

$$(A^{1/2})^2 = \begin{bmatrix} 4/3 & -1/3 \\ -2/3 & 5/3 \end{bmatrix} \begin{bmatrix} 4/3 & -1/3 \\ -2/3 & 5/3 \end{bmatrix} = \begin{bmatrix} 2 & -1 \\ -2 & 3 \end{bmatrix}$$

The reader may verify that the other square roots are:

$$A = \begin{bmatrix} 0 & 1 \\ 2 & -1 \end{bmatrix}$$

$$A = \begin{bmatrix} 0 & -1 \\ -2 & 1 \end{bmatrix}$$

and

$$A = \begin{bmatrix} -4/3 & 1/3 \\ 2/3 & -5/3 \end{bmatrix}$$

7.15 EIGENVALUES AND EIGENVECTORS VIA AN ITERATIVE PROCESS

A computationally efficient method is available for the determination of the eigenvalues and eigenvectors of higher-order square matrices. This method eliminates the tedious and error-prone hand calculations that are a necessary frustration to the successful conclusion of the process.

The largest eigenvalue and the corresponding eigenvector can be obtained by choosing an arbitrary $n \times 1$ eigenvector X, setting x_n equal to 1, and using it in Eq. (7.1).

Example: The symmetric matrix A

$$A = \begin{bmatrix} 4 & -2 & 0 \\ -2 & 3 & -1 \\ 0 & -1 & 2 \end{bmatrix}$$

is known (see the example in Section 7.11) to have eigenvalues of $\lambda_1 = 5.6691$, $\lambda_2 = 2.4760$, and $\lambda_3 = 0.8549$. The eigenvector corresponding to $\lambda_1 = 5.6691$ is:

$$X_1 = \alpha_1 \begin{bmatrix} -1.1983 \\ 1.0000 \\ -0.2725 \end{bmatrix}$$

An iterative procedure can be employed to confirm this.

Assume that

$$X_1 = \begin{bmatrix} 1.0000 \\ -1.0000 \\ 1.0000 \end{bmatrix}$$

The first iteration, using Eq. (7.1), $AX = \lambda X$, provides

$$\begin{bmatrix} 4 & -2 & 0 \\ -2 & 3 & -1 \\ 0 & -1 & 2 \end{bmatrix} \begin{bmatrix} 1.0000 \\ -1.0000 \\ 1.0000 \end{bmatrix} = \begin{bmatrix} 6.0000 \\ -6.0000 \\ 3.0000 \end{bmatrix} = 3.0000 \begin{bmatrix} 2.0000 \\ -2.0000 \\ 1.0000 \end{bmatrix}$$

The second iteration yields

$$\begin{bmatrix} 4 & -2 & 0 \\ -2 & 3 & -1 \\ 0 & -1 & 2 \end{bmatrix} \begin{bmatrix} 2.0000 \\ -2.0000 \\ 1.0000 \end{bmatrix} = \begin{bmatrix} 12.0000 \\ -11.0000 \\ 4.0000 \end{bmatrix} = 4.0000 \begin{bmatrix} 3.0000 \\ -2.7500 \\ 1.0000 \end{bmatrix}$$

and then a third gives

$$\begin{bmatrix} 4 & -2 & 0 \\ -2 & 3 & -1 \\ 0 & -1 & 2 \end{bmatrix} \begin{bmatrix} 3.0000 \\ -2.7500 \\ 1.0000 \end{bmatrix} = \begin{bmatrix} 17.5000 \\ -15.2500 \\ 4.7500 \end{bmatrix} = 4.7500 \begin{bmatrix} 3.6842 \\ -3.2105 \\ 1.0000 \end{bmatrix}$$

The procedure continues until the end of the sixteenth iteration, where it is observed that $\lambda_1 = 5.6691$ which is the result obtained in the example of Section 7.11. The corresponding eigenvector is

$$X_1 = \alpha \begin{bmatrix} 4.3965 \\ -3.6691 \\ 1.0000 \end{bmatrix}$$

which differs from the eigenvalue obtained in Section 7.11:

$$X_1 = \alpha_1 \begin{bmatrix} -1.1983 \\ 1.0000 \\ -0.2725 \end{bmatrix}$$

but merely indicates that the arbitrary α is related to the equally arbitrary α_1 by $\alpha = \alpha_1/3.6697$.

The next-largest eigenvalue and corresponding eigenvector can be determined by exploiting the orthogonality feature of the eigenvectors of a symmetric matrix. Let the eigenvector corresponding to the largest eigenvalue already determined be designated as X with elements $x_1, x_2, x_3, \cdots, x_n$. Suppose the eigenvector corresponding to the next largest eigenvector is designated a Y with elements $y_1, y_2, y_3, \cdots, y_n$. Then because the eigenvectors X and Y are orthogonal,

$$X^TY = 0$$

and

$$x_1y_1 + x_2y_2 + x_3y_3 + \cdots + x_ny_n = 0$$

If this is solved for y_n, the result is:

$$y_n = -\frac{x_1}{x_n}y_1 - \frac{x_2}{x_n}y_2 - \frac{x_3}{x_n}y_3 \cdots - \frac{x_{n-1}}{x_n}y_{n-1}$$

With this result substituted into

$$[A - \lambda I]Y = 0$$

a system of n homogeneous equations in $n - 1$ unknowns results. Because the nth equation is redundant, it may be eliminated. The result is a new matrix B that is $n - 1 \times n - 1$ and has a largest eigenvalue of

$$[B - \mu I]Y = 0$$

where μ and Y may be found by the iterative process.

Example: Consider A,

$$A = \begin{bmatrix} 4 & -2 & 0 \\ -2 & 3 & -1 \\ 0 & -1 & 2 \end{bmatrix}$$

which is known to have an eigenvalue $\lambda_1 = 5.6691$ and a corresponding eigenvector

$$X = \begin{bmatrix} 4.3695 \\ -3.6691 \\ 1.0000 \end{bmatrix}$$

The orthogonality requirement says that another 3×1 eigenvector Y must exist such that

$$y_3 = -\frac{x_1}{x_3}y_1 - \frac{x_2}{x_3}y_2 = -4.3965y_1 + 3.6691y_2$$

But Y with eigenvalue μ must also satisfy $[A - \mu I]Y$ so that

$$(4 - \mu)y_1 - 2y_2 = 0$$

$$-2y_1 + (3 - \mu)y_2 - y_3 = 0$$

$$-y_2 + (2 - \mu)y_3 = 0$$

Because the third equation is redundant, it may be eliminated, leaving

$$(4 - \mu)y_1 - 2y_2 = 0$$

$$-2y_1 + (3 - \mu)y_2 - (-4.3965y_1 + 3.6691y_2) = 0$$

or

$$(4 - \mu)y_1 - 2y_2 = 0$$

$$2.3965y_1 + (-0.6691 - \mu)y_2 = 0$$

This provides the matrix form

$$\begin{bmatrix} 4 & -2 \\ 2.3965 & -0.6991 \end{bmatrix}\begin{bmatrix} y_1 \\ y_2 \end{bmatrix} = \mu\begin{bmatrix} y_1 \\ y_2 \end{bmatrix}$$

and the iteration process is ready to begin.

Assume

$$Y = \begin{bmatrix} 1.0000 \\ 1.0000 \end{bmatrix}$$

and obtain for the first iteration

$$\begin{bmatrix} 4 & -2 \\ 2.3965 & -0.6691 \end{bmatrix}\begin{bmatrix} 1.0000 \\ 1.0000 \end{bmatrix} = \begin{bmatrix} 2.0000 \\ 1.7274 \end{bmatrix} = 1.7274\begin{bmatrix} 1.1578 \\ 1.0000 \end{bmatrix}$$

Then for the second iteration,

$$\begin{bmatrix} 4 & -2 \\ 2.3965 & -0.6691 \end{bmatrix}\begin{bmatrix} 1.1578 \\ 1.0000 \end{bmatrix} = \begin{bmatrix} 2.6311 \\ 2.1056 \end{bmatrix} = 2.1056\begin{bmatrix} 1.2496 \\ 1.0000 \end{bmatrix}$$

and for the third

$$\begin{bmatrix} 4 & -2 \\ 2.3965 & -0.6691 \end{bmatrix}\begin{bmatrix} 1.2496 \\ 1.0000 \end{bmatrix} = \begin{bmatrix} 2.9984 \\ 2.3256 \end{bmatrix} = 2.3256\begin{bmatrix} 1.2893 \\ 1.0000 \end{bmatrix}$$

This procedure continues until the end of the eleventh iteration, where it is then noted that $\lambda_2 = \mu = 2.4761$ and

$$Y = \begin{bmatrix} 1.3124 \\ 1.0000 \end{bmatrix}$$

and with $y_3 = -4.3965y_1 + 3.6691y_2$, the second eigenvector for the matrix A becomes:

$$X_2 = \alpha_2\begin{bmatrix} 1.3124 \\ 1.0000 \\ -2.1007 \end{bmatrix}$$

and this agrees with the result obtained in the example in Section 7.11.

The procedure may be continued until all eigenvalues and eigenvectors are determined. An alternate procedure, however, is available which depends on the difficulty of evaluating the inverse of A. As indicated in Section 7.12, the smallest eigenvalue of A is the largest eigenvalue of A^{-1}.

Example: It is desired to find the smallest eigenvalue and eigenvector of A:

$$A = \begin{bmatrix} 4 & -2 & 0 \\ -2 & 3 & -1 \\ 0 & -1 & 2 \end{bmatrix}$$

The inverse of A is easy to obtain:

$$A^{-1} = \begin{bmatrix} 0.4167 & 0.3333 & 0.1667 \\ 0.3333 & 0.6667 & 0.3333 \\ 0.1667 & 0.3333 & 0.6667 \end{bmatrix}$$

and now the largest eigenvalue of A^{-1} (the smallest eigenvalue of A) along with the associated eigenvector can be determined via the iteration procedure. Assume that the eigenvector is:

$$X = \begin{bmatrix} 1.0000 \\ 1.0000 \\ 1.0000 \end{bmatrix}$$

Then the first iteration yields:

$$\begin{bmatrix} 0.4167 & 0.3333 & 0.1667 \\ 0.3333 & 0.6667 & 0.3333 \\ 0.1667 & 0.3333 & 0.6667 \end{bmatrix} \begin{bmatrix} 1.0000 \\ 1.0000 \\ 1.0000 \end{bmatrix} = \begin{bmatrix} 0.9167 \\ 1.3333 \\ 1.6667 \end{bmatrix} = 1.6667 \begin{bmatrix} 0.7587 \\ 1.1429 \\ 1.0000 \end{bmatrix}$$

Then for the second iteration:

$$\begin{bmatrix} 0.4167 & 0.3333 & 0.1667 \\ 0.3333 & 0.6667 & 0.3333 \\ 0.1667 & 0.3333 & 0.6667 \end{bmatrix} \begin{bmatrix} 0.7587 \\ 1.1429 \\ 1.0000 \end{bmatrix} = \begin{bmatrix} 0.8750 \\ 1.3571 \\ 1.1786 \end{bmatrix} = 1.1786 \begin{bmatrix} 0.7424 \\ 1.1515 \\ 1.0000 \end{bmatrix}$$

and for the third:

$$\begin{bmatrix} 0.4167 & 0.3333 & 0.1667 \\ 0.3333 & 0.6667 & 0.3333 \\ 0.1667 & 0.3333 & 0.6667 \end{bmatrix} \begin{bmatrix} 0.7424 \\ 1.1515 \\ 1.0000 \end{bmatrix} = \begin{bmatrix} 0.8598 \\ 1.3485 \\ 1.1742 \end{bmatrix} = 1.1742 \begin{bmatrix} 0.7323 \\ 1.1484 \\ 1.0000 \end{bmatrix}$$

This procedure continues until the end of the eighth iteration, where it is noted that $\lambda = 1.1698$ and

$$X = \alpha \begin{bmatrix} 0.7282 \\ 1.1451 \\ 1.0000 \end{bmatrix}$$

The foregoing value, $\lambda = 1.1698$, applies to the inverse of A. The corresponding eigenvalue of A is $\lambda_3 = 1/1.1698 = 0.8549$. The foregoing eigenvector is seen to be equal to the eigenvector obtained in the example of Section 7.11:

$$X = \alpha_3 \begin{bmatrix} 0.6359 \\ 1.0000 \\ 0.8733 \end{bmatrix}$$

if $\alpha_3 = 0.8733\alpha$.

7.16 HOTELLING'S DEFLATION

Hornbeck† points out that there are a number of strategies that may be used to obtain the eigenvalues and eigenvectors between the largest and the smallest. One of these is known as *Hotelling's deflation* in which the largest eigenvalue is extracted from the matrix in successive stages.

Let λ_1 be the largest eigenvalue and X_1 be the largest eigenvector of the matrix A. A deflated matrix D may be formed from

$$D = A - \frac{\lambda_1(X_1 X_1^T)}{X_1^T X_1} \tag{7.18}$$

and the process may be repeated as far as desired or, as Hornbeck points out, until it is overwhelmed by roundoff error.

7.17 EXERCISES

Exercises 7.1 through 7.5 pertain to the matrix A:

$$A = \begin{bmatrix} 4 & -1 & 0 & -1 \\ -1 & 3 & -1 & 0 \\ 0 & -1 & 3 & -1 \\ -1 & 0 & -1 & 2 \end{bmatrix}$$

†Hornbeck, R. W., *Numerical Methods*, Quantum Publishers, New York, 1975.

7.1 Find the characteristic equation by expanding the characteristic determinant.

7.2 Find the characteristic equation by employing Bocher's method.

7.3 Find the characteristic equation by employing Leverrier's method.

7.4 Use Leverrier's method to find the inverse of A.

7.5 What are the eigenvalues of A?

Exercises 7.6 through 7.15 pertain to the matrix B:

$$B = \begin{bmatrix} 0 & 0 & -2 \\ 0 & -2 & 0 \\ -2 & 0 & 3 \end{bmatrix}$$

7.6 Find the characteristic equation by expanding the characteristic determinant.

7.7 Use Bocher's method to find the characteristic equation.

7.8 Use Leverrier's method to find the characteristic equation.

7.9 Use Leverrier's method to find the inverse of B.

7.10 What are the eigenvalues of B?

7.11 What are the eigenvectors of B?

7.12 What is the modal matrix of B?

7.13 Diagonalize B.

7.14 Determine B^5.

7.15 Use the iteration procedure to obtain the eigenvalues and eigenvectors of C:

$$C = \begin{bmatrix} 3 & -1 & -1 \\ -1 & 4 & -2 \\ -1 & -2 & 4 \end{bmatrix}$$

EIGHT

MATRIX POLYNOMIALS AND THE CALCULUS OF MATRICES

8.1 INTRODUCTION

An mth-order polynomial may be represented as

$$p(x) = \alpha_m x^m + \alpha_{m-1} x^{m-1} + \alpha_{m-2} x^{m-2} + \cdots + \alpha_2 x^2 + \alpha_1 x + \alpha_0$$

Although there are some notable differences between the algebra of scalars and the algebra of matrices, there are many similarities. For example, it has been observed that if A is a square matrix it may be raised to a power such as

$$A^5 = AAAAA$$

and this leads to considerations of adding exponents, as in

$$A^p A^q = A^{p+q}$$

Moreover, the multiplication of A by itself, no matter how many times, is certainly commutative, and if only A is involved, multiplication is distributive, as in

$$A(A + A) = AA + AA = 2AA = 2A^2$$

It would seem, therefore, that a matrix polynomial could be defined in the same manner as an algebraic polynomial. This is done in this chapter, as is a study of an infinite series of matrices and matrix functions.

8.2 THE CAYLEY–HAMILTON THEOREM

The *Cayley–Hamilton theorem* states that a matrix satisfies its own characteristic equation. The proof provides some introductory insight regarding the formulation and manipulation of matrix polynomials.

The characteristic matrix of A has been defined in Eq. (7.4):

$$K = [A - \lambda I]$$

and this matrix possesses an adjoint $B = \text{adj } K$ which is the transpose of the cofactor matrix $B = (K^c)$. Because A is $n \times n$, the elements of K^c and of adj K possess a maximum degree of $n - 1$.

Example: The 3×3 matrix:

$$A = \begin{bmatrix} 0 & 1 & 0 \\ 0 & 0 & 1 \\ -6 & -11 & -6 \end{bmatrix}$$

has a characteristic matrix:

$$K = [A - \lambda I] = \begin{bmatrix} -\lambda & 1 & 0 \\ 0 & -\lambda & 1 \\ -6 & -11 & -(6 + \lambda) \end{bmatrix}$$

and nine cofactors:

$$K_{11} = + \begin{vmatrix} -\lambda & 1 \\ -11 & -(6 + \lambda) \end{vmatrix} = \lambda^2 + 6\lambda + 11$$

$$K_{12} = - \begin{vmatrix} 0 & 1 \\ -6 & -(6 + \lambda) \end{vmatrix} = -6$$

$$K_{13} = + \begin{vmatrix} 0 & -\lambda \\ -6 & -11 \end{vmatrix} = -6\lambda$$

$$K_{21} = - \begin{vmatrix} 1 & 0 \\ -11 & -(6 + \lambda) \end{vmatrix} = \lambda + 6$$

$$K_{22} = + \begin{vmatrix} -\lambda & 0 \\ -6 & -(6 + \lambda) \end{vmatrix} = \lambda^2 + 6\lambda$$

$$K_{23} = - \begin{vmatrix} -\lambda & 1 \\ -6 & -11 \end{vmatrix} = -11\lambda - 6$$

$$K_{31} = + \begin{vmatrix} 1 & 0 \\ -\lambda & 1 \end{vmatrix} = 1$$

$$K_{32} = - \begin{vmatrix} -\lambda & 0 \\ 0 & 1 \end{vmatrix} = \lambda$$

and

$$K_{33} = + \begin{vmatrix} -\lambda & 1 \\ 0 & -\lambda \end{vmatrix} = \lambda^2$$

The cofactor matrix is, therefore,

$$K^c = \begin{bmatrix} (\lambda^2 + 6\lambda + 11) & -6 & -6\lambda \\ (\lambda + 6) & (\lambda^2 + 6\lambda) & -(11\lambda + 6) \\ 1 & \lambda & \lambda^2 \end{bmatrix}$$

and it can be noted that no element in K possesses a degree higher than $n - 1 = 3 - 1 = 2$. The adjoint of K is the transpose of K^c.

$$\text{adj } K = \begin{bmatrix} (\lambda^2 + 6\lambda + 11) & (\lambda + 6) & 1 \\ -6 & (\lambda^2 + 6\lambda) & \lambda \\ -6\lambda & -(11\lambda + 6) & \lambda^2 \end{bmatrix}$$

and this may be written as a polynomial:

$$B = B_2\lambda^2 + B_1\lambda + B_0$$

where B_2, B_1, and B_0 are 3×3 matrices. Here

$$B = \begin{bmatrix} 1 & 0 & 0 \\ 0 & 1 & 0 \\ 0 & 0 & 1 \end{bmatrix} \lambda^2 + \begin{bmatrix} 6 & 1 & 0 \\ 0 & 6 & 1 \\ -6 & -11 & 0 \end{bmatrix} \lambda + \begin{bmatrix} 11 & 6 & 1 \\ -6 & 0 & 0 \\ 0 & -6 & 0 \end{bmatrix}$$

It may be recalled (see Sec. 3.5) that

$$K \frac{\text{adj } K}{\det K} = I$$

With $K = (A - \lambda I)$, this can be rearranged to the form

$$[A - \lambda I]B = \det [A - \lambda I]$$

and because $\det [A - \lambda I]$ yields the characteristic polynomial, it is observed that

$$AB - \lambda B = p(\lambda)I \tag{8.1}$$

where λ is an eigenvalue of A. Here $p(\lambda)$ can be represented by

$$p(\lambda) = \alpha_n\lambda^n + \alpha_{n-1}\lambda^{n-1} + \alpha_{n-2}\lambda^{n-2} + \cdots + \alpha_2\lambda^2 + \alpha_1\lambda + \alpha_0 \quad (8.2)$$

or

$$p(\lambda) = \sum_{k=0}^{n} \alpha_k\lambda^k \quad (8.3)$$

It was noted in the example that the adjoint of K ($B = \text{adj } K$) can be written as a polynomial:

$$B = B_0 + B_1\lambda + B_2\lambda^2 + \cdots + B_{n-1}\lambda^{n-1} \quad (8.4)$$

and with this and Eq. (8.3) inserted into Eq. (8.1), the result is:

$$\sum_{k=0}^{n-1} AB_k\lambda^k - \sum_{k=0}^{n-1} B_k\lambda^{k+1} = \sum_{k=0}^{n} \alpha_k I\lambda^k \quad (8.5)$$

The coefficients of corresponding powers of λ in Eq. (8.5) may be equated so that a set of $n + 1$ equations relating the matrices A, B, and I and the coefficients α_k can be obtained:

$$\begin{aligned}
\lambda^0: &\quad AB_0 && = \alpha_0 I \\
\lambda^1: &\quad AB_1 - B_0 && = \alpha_1 I \\
\lambda^2: &\quad AB_2 - B_1 && = \alpha_2 I \\
&\quad \cdots \\
\lambda^{n-1}: &\quad AB_{n-1} - B_{n-2} && = \alpha_{n-1} I \\
\lambda^n: &\quad -B_{n-1} && = \alpha_n I
\end{aligned}$$

If the second of these is multiplied by A, the third by A^2, and so on up to the $(n + 1)$th equation, which must be multiplied by A^n, the result is:

$$\begin{aligned}
\lambda^0: &\; AB_0 && = \alpha_0 I \\
\lambda^1: &\; A^2B_1 - AB_0 && = \alpha_1 A \\
\lambda^2: &\; A^3B_2 - A^2B_1 && = \alpha_2 A^2 \\
&\; \cdots \\
\lambda^{n-1}: &\; A^nB_{n-1} - A^{n-1}B_{n-2} && = \alpha_{n-1}A^{n-1} \\
\lambda^n: &\; -A^nB_{n-1} && = \alpha_n A^n
\end{aligned}$$

If these equations are added, a polynomial equation in the matrix A results:

$$p(A) = \alpha_n A^n + \alpha_{n-1}A^{n-1} + \alpha_{n-2}A^{n-2}$$

$$+ \cdots + \alpha_2 A^2 + \alpha_1 A + \alpha_0 I = 0 \qquad (8.6)$$

and if Eq. (8.2) is set equal to zero,

$$p(\lambda) = \alpha_n \lambda^n + \alpha_{n-1}\lambda^{n-1} + \alpha_{n-2}\lambda^{n-2} + \cdots + \alpha_2\lambda^2 + \alpha_1\lambda + \alpha_0 = 0$$

a comparison can be established that shows that the matrix A satisfies its own polynomial equation. This is the statement of the Cayley–Hamilton theorem.

8.3 TWO APPLICATIONS OF THE CAYLEY–HAMILTON THEOREM

Because a matrix must satisfy its own characteristic equation, Eq. (8.6) suggests that it is possible to determine A^n in terms of a polynomial in A whose degree does not exceed $n - 1$:

$$A^n = -\frac{1}{\alpha_n}(\alpha_{n-1}A^{n-1} + \alpha_{n-2}A^{n-2} + \alpha_{n-3}A^{n-3} + \cdots + \alpha_1 A + \alpha_0 I)$$

$$(8.7)$$

This fact greatly simplifies the evaluation of matrix polynomials.

Example: It is desired to evaluate $f(A) = A^4 + 8A$ with

$$A = \begin{bmatrix} 4 & -2 \\ -2 & 3 \end{bmatrix}$$

The characteristic equation of A is obtained from det $[A - \lambda I] = 0$:

$$\det \begin{bmatrix} (4 - \lambda) & -2 \\ -2 & (3 - \lambda) \end{bmatrix} = 0$$

so that

$$p(\lambda) = \lambda^2 - 7\lambda + 8 = 0$$

By invoking the Cayley–Hamilton theorem, one obtains

$$p(A) = A^2 - 7A + 8I = 0$$

Thus

$$A^2 = 7A - 8I$$

and

$$A^4 = (A^2)^2 = 49A^2 - 112A + 64I$$
$$= 49(7A - 8I) - 112A + 64I$$

or

$$A^4 = 231A - 328I$$

Thus the required $f(A) = A^4 + 8A$ is:

$$f(A) = A^4 + 8I = (231A - 328I) + 8A$$

or

$$f(A) = 239A - 328I = 239\begin{bmatrix} 4 & -2 \\ -2 & 3 \end{bmatrix} - 328\begin{bmatrix} 1 & 0 \\ 0 & 1 \end{bmatrix}$$

$$= \begin{bmatrix} 956 & -478 \\ -478 & 717 \end{bmatrix} - \begin{bmatrix} 328 & 0 \\ 0 & 328 \end{bmatrix}$$

and

$$f(A) = A^4 + 8I = 239A - 328I = \begin{bmatrix} 628 & -478 \\ -478 & 389 \end{bmatrix}$$

This result may be confirmed in a simpler, brute-force, long-winded method. With

$$A^2 = \begin{bmatrix} 4 & -2 \\ -2 & 3 \end{bmatrix}\begin{bmatrix} 4 & -2 \\ -2 & 3 \end{bmatrix} = \begin{bmatrix} 20 & -14 \\ -14 & 13 \end{bmatrix}$$

and then

$$A^4 = (A^2)^2 = \begin{bmatrix} 20 & -14 \\ -14 & 13 \end{bmatrix}\begin{bmatrix} 20 & -14 \\ -14 & 13 \end{bmatrix} = \begin{bmatrix} 596 & -462 \\ -462 & 365 \end{bmatrix}$$

$f(A) = A^4 + 8A$ is evaluated as

$$f(A) = A^4 + 8I = \begin{bmatrix} 596 & -462 \\ -462 & 365 \end{bmatrix} + 8\begin{bmatrix} 4 & -2 \\ -2 & 3 \end{bmatrix}$$

$$= \begin{bmatrix} 596 & -462 \\ -462 & 365 \end{bmatrix} + \begin{bmatrix} 32 & -16 \\ -16 & 24 \end{bmatrix} = \begin{bmatrix} 628 & -478 \\ -478 & 389 \end{bmatrix}$$

which confirms the previous result obtained via the Cayley–Hamilton theorem.

The Cayley–Hamilton theorem is useful for obtaining inverses. If Eq. (8.6) is multiplied by A^{-1}, then

$$A^{-1}p(A) = \alpha_n A^{n-1} + \alpha_{n-1}A^{n-2} + \alpha_{n-2}A^{n-3}$$
$$+ \cdots + \alpha_2 A + \alpha_1 I + \alpha_0 A^{-1} = 0$$

and this may be solved for the inverse of A

$$A^{-1} = -\frac{1}{\alpha_0}(\alpha_n A^{n-1} + \alpha_{n-1}A^{n-2}$$
$$+ \alpha_{n-2}A^{n-3} + \cdots + \alpha_2 A + \alpha_1 I) \qquad (8.8)$$

Observe that the inverse of A has been expressed in terms of powers of A that do not exceed $n - 1$.

Example: The inverse of the matrix A:

$$A = \begin{bmatrix} 3 & -2 & 0 & 0 \\ -2 & 4 & -1 & 0 \\ 0 & -1 & 6 & -1 \\ 0 & 0 & -1 & 3 \end{bmatrix}$$

can be determined using the Cayley–Hamilton theorem in a procedure summarized by Eq. (8.8).

The first step is to obtain the characteristic equation. Because A^2 and A^3 are required in any event, Bocher's method may be employed:

$$A^2 = \begin{bmatrix} 3 & -2 & 0 & 0 \\ -2 & 4 & -1 & 0 \\ 0 & -1 & 6 & -1 \\ 0 & 0 & -1 & 3 \end{bmatrix}\begin{bmatrix} 3 & -2 & 0 & 0 \\ -2 & 4 & -1 & 0 \\ 0 & -1 & 6 & -1 \\ 0 & 0 & -1 & 3 \end{bmatrix}$$

$$= \begin{bmatrix} 13 & -14 & 2 & 0 \\ -14 & 21 & -10 & 1 \\ 2 & -10 & 38 & -9 \\ 0 & 1 & -9 & 10 \end{bmatrix}$$

and

$$A^3 = AA^2 = \begin{bmatrix} 3 & -2 & 0 & 0 \\ -2 & 4 & -1 & 0 \\ 0 & -1 & 6 & -1 \\ 0 & 0 & -1 & 3 \end{bmatrix} \begin{bmatrix} 13 & -14 & 2 & 0 \\ -14 & 21 & -10 & 1 \\ 2 & -10 & 38 & -9 \\ 0 & 1 & -9 & 10 \end{bmatrix}$$

$$= \begin{bmatrix} 67 & -84 & 26 & -2 \\ -84 & 122 & -82 & 13 \\ 26 & -82 & 247 & -65 \\ -2 & 13 & -65 & 39 \end{bmatrix}$$

The traces are:

$$\text{tr } A = \mu_1 = 16$$

$$\text{tr } A^2 = \mu_2 = 82$$

and

$$\text{tr } A^3 = \mu_3 = 475$$

These lead to the evaluation of three of the five coefficients in the characteristic equation:

$$\beta_1 = -\mu_1 = -16$$

$$\beta_2 = -\frac{1}{2}(\mu_1\beta_1 + \mu_2) = -\frac{1}{2}(-256 + 82) = 87$$

and

$$\beta_3 = -\frac{1}{3}(\mu_1\beta_2 + \mu_2\beta_1 + \mu_3)$$

$$= -\frac{1}{3}[87(16) + (-16)(82) + 475]$$

or

$$\beta_3 = -555/3 = -185$$

and it remains to evaluate β_4, which can be determined by either squaring A^2 or by recognizing (see Sec. 7.3) that $\det A = (-1)^n\beta_4$. Observe that A contains two zeroes in its first row (and its first column) and that it might well be easier to evaluate $\det A$ than to multiply two 4×4 matrices. Thus the procedure of finding det A is adopted and via the Laplace expansion,

$$\det \begin{bmatrix} 3 & -2 & 0 & 0 \\ -2 & 4 & -1 & 0 \\ 0 & -1 & 6 & -1 \\ 0 & 0 & -1 & 3 \end{bmatrix} = 3 \begin{vmatrix} 4 & -1 & 0 \\ -1 & 6 & -1 \\ 0 & -1 & 3 \end{vmatrix} + 2 \begin{vmatrix} -2 & -1 & 0 \\ 0 & 6 & -1 \\ 0 & -1 & 3 \end{vmatrix}$$

or det $A = 3(72 - 4 - 3) + 2(-36 + 2) = 3(65) + 2(-34) = 127$,
and with $n = 4$ so that $(-1)^4 = 1$, it is seen that $\beta_4 = \det A = 127$.

All the coefficients of the characteristic equation have now been established and the characteristic equation is:

$$p(\lambda) = \lambda^4 - 16\lambda^3 + 87\lambda^2 - 185\lambda + 127 = 0$$

and in terms of the matrix A and as a consequence of the Cayley–Hamilton theorem:

$$p(A) = A^4 - 16A^3 + 87A^2 - 185A + 127I = 0$$

Now, in accordance with Eq. (8.8), the sought-after inverse of A can be evaluated from

$$A^{-1} = -\frac{1}{127}(A^3 - 16A^2 + 87A - 185I)$$

The evaluation proceeds by letting $C_1 = A^3 - 16A^2$

$$C_1 = \begin{bmatrix} 67 & -84 & 26 & -2 \\ -84 & 122 & -82 & 13 \\ 26 & -82 & 247 & -65 \\ -2 & 13 & -65 & 39 \end{bmatrix} - 16 \begin{bmatrix} 13 & -14 & 2 & 0 \\ -14 & 21 & -10 & 1 \\ 2 & -10 & 38 & -9 \\ 0 & 1 & -9 & 10 \end{bmatrix}$$

$$= \begin{bmatrix} 67 & -84 & 26 & -2 \\ -84 & 122 & -82 & 13 \\ 26 & -82 & 247 & -65 \\ -2 & 13 & -65 & 39 \end{bmatrix} - \begin{bmatrix} 208 & -224 & 32 & 0 \\ -224 & 336 & -160 & 16 \\ 32 & -160 & 608 & -144 \\ 0 & 16 & -144 & 160 \end{bmatrix}$$

or

$$C_1 = \begin{bmatrix} -141 & 140 & -6 & -2 \\ 140 & -214 & 78 & -3 \\ -6 & 78 & -361 & 79 \\ -2 & -3 & 79 & -121 \end{bmatrix}$$

Then with $C_2 = 87A - 185I$,

$$C_2 = 87 \begin{bmatrix} 3 & -2 & 0 & 0 \\ -2 & 4 & -1 & 0 \\ 0 & -1 & 6 & -1 \\ 0 & 0 & -1 & 3 \end{bmatrix} - 185 \begin{bmatrix} 1 & 0 & 0 & 0 \\ 0 & 1 & 0 & 0 \\ 0 & 0 & 1 & 0 \\ 0 & 0 & 0 & 1 \end{bmatrix}$$

$$= \begin{bmatrix} 261 & -174 & 0 & 0 \\ -174 & 348 & -87 & 0 \\ 0 & -87 & 522 & -87 \\ 0 & 0 & -87 & 261 \end{bmatrix} - \begin{bmatrix} 185 & 0 & 0 & 0 \\ 0 & 185 & 0 & 0 \\ 0 & 0 & 185 & 0 \\ 0 & 0 & 0 & 185 \end{bmatrix}$$

or

$$C_2 = \begin{bmatrix} 76 & -174 & 0 & 0 \\ -174 & 163 & -87 & 0 \\ 0 & -87 & 337 & -87 \\ 0 & 0 & -87 & 76 \end{bmatrix}$$

Then

$$A = -\frac{1}{127}(C_1 + C_2)$$

or

$$A = -\frac{1}{127}\begin{bmatrix} -141 & 140 & -6 & -2 \\ 140 & -214 & 78 & -3 \\ -6 & 78 & -361 & 79 \\ -2 & -3 & 79 & -121 \end{bmatrix}$$

$$+ \begin{bmatrix} 76 & -174 & 0 & 0 \\ -174 & 163 & -87 & 0 \\ 0 & -87 & 337 & -87 \\ 0 & 0 & -87 & 76 \end{bmatrix}$$

and the final result is:

$$A = \frac{1}{127} \begin{bmatrix} 65 & 34 & 6 & 2 \\ 34 & 51 & 9 & 3 \\ 6 & 9 & 24 & 8 \\ 2 & 3 & 8 & 45 \end{bmatrix}$$

which the reader may wish to verify.

Before leaving this section it is well to remark that the method chosen to evaluate the inverse of the matrix A in the foregoing example is probably not the best available. It is questionable whether the labor involved in obtaining the coefficients of the characteristic equation is less than that required to obtain 16 cofactors or to determine the inverse by partitioning A into four 2×2 matrices. It is also to be noted that Leverrier's method can probably beat Bocher's method in a race for the coefficients of the characteristic equation and has the advantage that the inverse of A can be obtained directly without resorting to the employment of Eq. (8.8).

8.4 AN INTERESTING APPLICATION OF THE REMAINDER THEOREM OF ALGEBRA

The *remainder theorem of algebra* says that if a polynomial $f(x)$ is divided by another polynomial $p(x)$, the result is a quotient polynomial $q(x)$ plus a remainder $r(x)$ divided by $p(x)$. This means that

$$\frac{f(x)}{p(x)} = q(x) + \frac{r(x)}{p(x)}$$

and if this is multiplied by $p(x)$,

$$f(x) = q(x) \cdot p(x) + r(x)$$

If one considers $p(x)$ as a polynomial in λ, then

$$f(\lambda) = q(\lambda)p(\lambda) + r(\lambda)$$

and if $p(\lambda)$ is then chosen as the characteristic polynomial of the $n \times n$ matrix A, then, in accordance with the Cayley–Hamilton theorem, $p(\lambda) = p(A) = 0$ and

$$f(A) = r(A) \tag{8.9}$$

Equation (8.9) can be employed to evaluate functions of the matrix A.

Example: If

$$A = \begin{bmatrix} 4 & -2 \\ -2 & 3 \end{bmatrix}$$

it is known that

$$p(\lambda) = \lambda^2 - 7\lambda + 8$$

If it is desired to evaluate $f(A) = A^4 + 8A$ as in the first example in Section 8.3, the evaluation may be accomplished by finding $r(A)$.

Let

$$f(x) = x^4 + 8x$$

and

$$p(x) = x^2 - 7x + 8$$

Long division provides

$$\frac{f(x)}{p(x)} = x^2 + 7x + 41 \frac{239x - 328}{x^2 - 7x + 8}$$

and here it is observed that the remainder is

$$r(x) = 239x - 328$$

Thus, in accordance with Eq. (8.10), this makes

$$f(A) = r(A) = 239A - 328I$$

which agrees with the result obtained from a direct application of the Cayley–Hamilton theorem in one of the examples in Section 8.3.

8.5 INFINITE SERIES INVOLVING MATRICES

An infinite series in the real or complex variable z can be set down as

$$f(z) = \alpha_0 + \alpha_1 z + \alpha_2 z^2 + \alpha_3 z^3 + \cdots + \alpha_j z^j + \cdots$$

and this can be written as

$$f(z) = \sum_{k=0}^{\infty} \alpha_k z^k$$

If the real or complex variable z is replaced by the $n \times n$ matrix A, then

$$f(A) = \sum_{k=0}^{\infty} \alpha_k A^k \qquad (8.10)$$

and this is an infinite series in the matrix A. If a matrix S_j is defined as the sum of the first j terms of Eq. (8.10),

$$S_j = \alpha_0 I + \alpha_1 A + \alpha_2 A + \cdots + \alpha_{j-1} A^{j-1} + \alpha_j A^j$$

then the infinite series of Eq. (8.10) converges if there exists a matrix S with bounded elements (a sum) such that

$$S = \lim_{j \to \infty} S_j$$

Now it is shown in elementary calculus that if z is a real or complex variable, the infinite series such as e^z, $\cos z$, $\sin z$, $\cosh z$ and $\sinh z$ all converge. This means that

$$e = I + \frac{A}{1!} + \frac{A^2}{2!} + \frac{A^3}{3!} + \cdots$$

$$\cos A = I - \frac{A^2}{2!} + \frac{A^4}{4!} + \frac{A^6}{6!} + \cdots$$

$$\sin A = \frac{A}{1!} - \frac{A^3}{3!} + \frac{A^5}{5!} - \frac{A^7}{7!} + \cdots$$

$$\cosh A = I + \frac{A^2}{2!} + \frac{A^4}{4!} + \frac{A^6}{6!} + \cdots$$

and

$$\sinh A = \frac{A}{1!} + \frac{A^3}{3!} + \frac{A^5}{5!} + \frac{A^7}{7!} + \cdots$$

This also leads to the familiar Euler relationships:

$$e^{jA} = \cos A + j \sin A$$

and

$$e^{-jA} = \cos A - j \sin A$$

8.6 SYLVESTER'S THEOREM

Sylvester's theorem can be used to determine a function of the $n \times n$ matrix A. The proof requires the employment of contour integrals in the complex plane, and, because a demonstration of the proof serves no useful purpose, it is omitted here.

Sylvester's theorem (often called *Sylvester's identity*) is

$$f(A) = \sum_{k=1}^{n} f(\lambda_k) \frac{\prod_{\substack{i=1 \\ i \neq k}}^{n} A - \lambda_i I}{\prod_{\substack{i=1 \\ i \neq k}}^{n} (\lambda_k - \lambda_i)} \qquad (8.11)$$

Care should be exercised in handling the products indicated by the large *pi*'s. Each of these involve a product with *i* as the index but where *i* can never equal *k*.

The use of Sylvester's theorem is best illustrated by an example.

Example: It is desired to find e^{At} if A is

$$A = \begin{bmatrix} 0 & 1 & 0 \\ 0 & 0 & 1 \\ -6 & -11 & -6 \end{bmatrix}$$

It is known (see Section 7.6) that the matrix A possesses a characteristic polynomial

$$p(\lambda) = \lambda^3 + 6\lambda^2 + 11\lambda + 6$$

and this leads to the three eigenvalues of A:

$$\lambda_1 = -1$$

$$\lambda_2 = -2$$

and

$$\lambda_3 = -3$$

The use of Sylvester's theorem for this 3 × 3 matrix calls for three terms in a summation. For $\lambda_1 = -1$,

$$(A - \lambda_2 I)(A - \lambda_3 I) = A^2 - (\lambda_2 + \lambda_3)A + \lambda_2\lambda_3 I$$

or with $\lambda_2 = -2$ and $\lambda_3 = -3$, $A^2 + 5A + 6I$ becomes

$$\begin{bmatrix} 0 & 1 & 0 \\ 0 & 0 & 1 \\ -6 & -11 & -6 \end{bmatrix}\begin{bmatrix} 0 & 1 & 0 \\ 0 & 0 & 1 \\ -6 & -11 & -6 \end{bmatrix}$$

$$+ 5\begin{bmatrix} 0 & 1 & 0 \\ 0 & 0 & 1 \\ -6 & -11 & -6 \end{bmatrix} + 6\begin{bmatrix} 1 & 0 & 0 \\ 0 & 1 & 0 \\ 0 & 0 & 1 \end{bmatrix}$$

or

$$
\begin{bmatrix} 0 & 0 & 1 \\ -6 & -11 & -6 \\ 36 & 60 & 25 \end{bmatrix} + \begin{bmatrix} 0 & 5 & 0 \\ 0 & 0 & 5 \\ -30 & -55 & -30 \end{bmatrix} + \begin{bmatrix} 6 & 0 & 0 \\ 0 & 6 & 0 \\ 0 & 0 & 6 \end{bmatrix}
$$

$$
= \begin{bmatrix} 6 & 5 & 1 \\ -6 & -5 & -1 \\ 6 & 5 & 1 \end{bmatrix}
$$

and with $[-1 - (-2)][-1 - (-3)] = (1)(2) = 2$, the first term is:

$$
\text{term}_1 = \frac{e^{-t}}{2} \begin{bmatrix} 6 & 5 & 1 \\ -6 & -5 & -1 \\ 6 & 5 & 1 \end{bmatrix} = \begin{bmatrix} 3e^{-t} & 5/2e^{-t} & 1/2e^{-t} \\ -3e^{-t} & -5/2e^{-t} & -1/2e^{-t} \\ 3e^{-t} & 5/2e^{-t} & 1/2e^{-t} \end{bmatrix}
$$

For $\lambda_2 = -2$,

$$
(A + I)(A + 3I) = A^2 + 4A + 3I
$$

or

$$
\begin{bmatrix} 0 & 0 & 1 \\ -6 & -11 & -6 \\ 36 & 60 & 25 \end{bmatrix} + \begin{bmatrix} 0 & 4 & 0 \\ 0 & 0 & 4 \\ -24 & -44 & -24 \end{bmatrix} + \begin{bmatrix} 3 & 0 & 0 \\ 0 & 3 & 0 \\ 0 & 0 & 3 \end{bmatrix}
$$

$$
= \begin{bmatrix} 3 & 4 & 1 \\ -6 & -8 & -2 \\ 12 & 16 & 4 \end{bmatrix}
$$

and with $[-2 - (-1)][-2 - (-3)] = (-1)(1) = -1$, the second term is:

$$
\text{term}_2 = \frac{e^{-2t}}{-1} \begin{bmatrix} 3 & 4 & 1 \\ -6 & -8 & -2 \\ 12 & 16 & 4 \end{bmatrix} = \begin{bmatrix} -3e^{-2t} & -4e^{-2t} & -e^{-2t} \\ 6e^{-2t} & 8e^{-2t} & 2e^{-2t} \\ -12e^{-2t} & -16e^{-2t} & -4e^{-2t} \end{bmatrix}
$$

Finally, for $\lambda_3 = -3$,

$$
(A + I)(A + 2I) = A^2 + 3A + 2I
$$

or

$$
\begin{bmatrix} 0 & 0 & 1 \\ -6 & -11 & -6 \\ 36 & 60 & 25 \end{bmatrix} + \begin{bmatrix} 0 & 3 & 0 \\ 0 & 0 & 3 \\ -18 & -33 & -18 \end{bmatrix} + \begin{bmatrix} 2 & 0 & 0 \\ 0 & 2 & 0 \\ 0 & 0 & 2 \end{bmatrix}
$$

$$
= \begin{bmatrix} 2 & 3 & 1 \\ -6 & -9 & -3 \\ 18 & 27 & 9 \end{bmatrix}
$$

and with $[-3 -(-1)][-3 -(-2)] = (-2)(-1) = 2$, the third term is:

$$
\text{term}_3 = \frac{e^{-3t}}{2} \begin{bmatrix} 2 & 3 & 1 \\ -6 & -9 & -3 \\ 18 & 27 & 9 \end{bmatrix} = \begin{bmatrix} e^{-3t} & 3/2e^{-3t} & 1/2e^{-3t} \\ -3e^{-3t} & -9/2e^{-3t} & -3/2e^{-3t} \\ 9e^{-3t} & 27/2e^{-3t} & 9/2e^{-3t} \end{bmatrix}
$$

Thus

$$
f(A) = e^{At} = \text{term}_1 + \text{term}_2 + \text{term}_3
$$

and

$$
e^{At} = \frac{e^{-t}}{2} \begin{bmatrix} 6 & 5 & 1 \\ -6 & -5 & -1 \\ 6 & 5 & 1 \end{bmatrix} - \frac{e^{-2t}}{1} \begin{bmatrix} 3 & 4 & 1 \\ -6 & -8 & -2 \\ 12 & 16 & 4 \end{bmatrix}
$$

$$
+ \frac{e^{-3t}}{2} \begin{bmatrix} 2 & 3 & 1 \\ -6 & -9 & -3 \\ 18 & 27 & 9 \end{bmatrix}
$$

and this may be written as

$$
e^{At} = \begin{bmatrix} (\ e^{-3t} & -3e^{-2t} & +3e^{-t}) \\ (-3e^{-3t} & +6e^{-2t} & -3e^{-t}) \\ (\ 9e^{-3t} & -12e^{-2t} & +3e^{-t}) \end{bmatrix}
$$

$$
\begin{bmatrix} (\ \tfrac{3}{2}e^{-3t} & -4e^{-2t} & +\tfrac{5}{2}e^{-t}) & (\ \tfrac{1}{2}e^{-3t} & -e^{-2t} & +\tfrac{1}{2}e^{-t}) \\ (-\tfrac{9}{2}e^{-3t} & +8e^{-2t} & -\tfrac{5}{2}e^{-t}) & (-\tfrac{3}{2}e^{-3t} & +2e^{-2t} & -\tfrac{1}{2}e^{-t}) \\ (\ \tfrac{27}{2}e^{-3t} & -16e^{-2t} & +\tfrac{5}{2}e^{-t}) & (\ \tfrac{9}{2}e^{-3t} & -4e^{-2t} & +\tfrac{1}{2}e^{-t}) \end{bmatrix}
$$

Example: This time it is desired to find cos At with

$$A = \begin{bmatrix} 1 & -3 \\ -1 & 2 \end{bmatrix}$$

Here the characteristic matrix is:

$$K = [A - \lambda I] = \begin{bmatrix} (4 - \lambda) & -3 \\ -1 & (2 - \lambda) \end{bmatrix}$$

and this leads to the characteristic polynomial

$$\det K = \lambda^2 - 6\lambda + 5$$

and two eigenvalues

$$\lambda_1 = 5$$

and

$$\lambda_2 = 1$$

For $f(A) = \cos At$

$$f(\lambda_1 = 5) = \cos 5t$$

and

$$f(\lambda_2 = 1) = \cos t$$

Then, with $\lambda_1 = 5$,

$$\lambda_1 - \lambda_2 = 5 - 1 = 4$$

and

$$[A - I] = \begin{bmatrix} 3 & -3 \\ -1 & 1 \end{bmatrix}$$

one term of cos At becomes

$$term_1 = \frac{\cos 5t}{4} \begin{bmatrix} 3 & -3 \\ -1 & 1 \end{bmatrix}$$

For $\lambda_2 = 1$,

$$\lambda_2 - \lambda_1 = 1 - 5 = -4$$

and

$$A - 5I = \begin{bmatrix} -1 & -3 \\ -1 & -3 \end{bmatrix}$$

The other term of cos At is therefore

$$\text{term}_2 = \frac{\cos t}{-4} \begin{bmatrix} -1 & -3 \\ -1 & -3 \end{bmatrix}$$

The sought-after result is obtained from Sylvester's theorem. Adding term$_1$ to term$_2$ yields:

$$\cos At = \frac{\cos 5t}{4} \begin{bmatrix} 3 & -3 \\ -1 & 1 \end{bmatrix} + \frac{\cos t}{-4} \begin{bmatrix} -1 & -3 \\ -1 & -3 \end{bmatrix}$$

or

$$\cos At = \frac{1}{4} \begin{bmatrix} (3\cos 5t + \cos t) - & (3\cos 5t - 3\cos t) \\ -(\cos 5t - \cos t) & (\cos 5t + 3\cos t) \end{bmatrix}$$

8.7 MATRIX FUNCTIONS WITH DISTINCT EIGENVALUES

Suppose that in

$$f(x) = q(x)\, p(x) + r(x)$$

where x is set equal to λ so that $f(x)$ is an infinite series in λ, it is required to evaluate a matrix function $f(A)$. In this case, $p(\lambda)$ becomes the characteristic polynomial of the matrix A and $q(x)$ is a polynomial in x that is also some function of λ. The polynomial $r(x)$ is the remainder (see Sec. 8.4). It is also a function of λ, but if A is $n \times n$, $r(\lambda)$ must be a polynomial of degree $n - 1$ so that in

$$f(\lambda) = q(\lambda)\, p(\lambda) + r(\lambda) \tag{8.12}$$

$r(\lambda)$ looks like

$$r(\lambda) = \alpha_0 + \alpha_1\lambda + \alpha_2\lambda^2 + \cdots + \alpha_{n-2}\lambda^{n-2} + \alpha_{n-1}\lambda^{n-1} \tag{8.13}$$

The coefficients of $r(\lambda)$ may be obtained by exploiting the fact that λ is a root of $p(\lambda) = 0$ so that $p(\lambda) = 0$ for $k = 1, 2, 3, \cdots, n$. Thus

$$f(\lambda_1) = r(\lambda_1)$$
$$f(\lambda_2) = r(\lambda_2)$$
$$f(\lambda_3) = r(\lambda_3)$$
$$\cdots$$
$$f(\lambda_n) = r(\lambda_n)$$

represents a set of n algebraic equations in the n coefficients of $r(\lambda)$. Solution of this set determines the α's in Eq. (8.13). Moreover, Eq. (8.9) points out that because $p(A) = 0$, $f(A) = r(A)$.

Example: It is desired to find e^{At} if

$$A = \begin{bmatrix} 0 & 1 & 0 \\ 0 & 0 & 1 \\ -6 & -11 & -6 \end{bmatrix}$$

This example is done using the Sylvester theorem in Section 8.6.
The characteristic equation here is known to be:

$$p(\lambda) = \lambda^3 + 6\lambda^2 + 11\lambda + 6 = 0$$

and the eigenvalues are:

$$\lambda_1 = -1$$

$$\lambda_2 = -2$$

and

$$\lambda_3 = -3$$

In this case, $n = 3$ and for $f(A) = e^{At}$, $r(\lambda)$ is:

$$r(\lambda) = \alpha_0 + \alpha_1\lambda + \alpha_2\lambda^2$$

Use of the three eigenvalues provides:

$$\alpha_0 - \alpha_1 + \alpha_2 = e^{-t}$$

$$\alpha_0 - 2\alpha_1 + 4\alpha_2 = e^{-2t}$$

$$\alpha_0 - 3\alpha_2 + 9\alpha_2 = e^{-3t}$$

and this can be put into matrix form:

$$\begin{bmatrix} 1 & -1 & 1 \\ 1 & -2 & 4 \\ 1 & -3 & 9 \end{bmatrix} \begin{bmatrix} \alpha_0 \\ \alpha_1 \\ \alpha_2 \end{bmatrix} = \begin{bmatrix} e^{-t} \\ e^{-2t} \\ e^{-3t} \end{bmatrix}$$

A matrix inversion can be used to provide the solution

$$\begin{bmatrix} \alpha_0 \\ \alpha_1 \\ \alpha_2 \end{bmatrix} = \begin{bmatrix} 3 & -3 & 1 \\ 5/2 & -4 & 3/2 \\ 1/2 & -1 & 1/2 \end{bmatrix} \begin{bmatrix} e^{-t} \\ e^{-2t} \\ e^{-3t} \end{bmatrix}$$

or

$$\begin{bmatrix} \alpha_0 \\ \alpha_1 \\ \alpha_2 \end{bmatrix} = \begin{bmatrix} (3e^{-t} & -3e^{-2t} & + e^{-3t}) \\ (\tfrac{5}{2}e^{-t} & -4e^{-2t} & +\tfrac{3}{2}e^{-3t}) \\ (\tfrac{1}{2}e^{-t} & -e^{-2t} & +\tfrac{1}{2}e^{-3t}) \end{bmatrix}$$

With A^2 evaluated as

$$A^2 = \begin{bmatrix} 0 & 1 & 0 \\ 0 & 0 & 1 \\ -6 & -11 & -6 \end{bmatrix}\begin{bmatrix} 0 & 1 & 0 \\ 0 & 0 & 1 \\ -6 & -11 & -6 \end{bmatrix} = \begin{bmatrix} 0 & 0 & 1 \\ -6 & -11 & -6 \\ 36 & 60 & 25 \end{bmatrix}$$

and in accordance with Eq. (8.9):

$$f(A) = \alpha_0 I + \alpha_1 A + \alpha_2 A^2$$

or

$$e^{At} = (3e^{-t} - 3e^{-2t} + e^{-3t})\begin{bmatrix} 1 & 0 & 0 \\ 0 & 1 & 0 \\ 0 & 0 & 1 \end{bmatrix}$$

$$+ \left(\frac{5}{2}e^{-t} - 4e^{-2t} + \frac{3}{2}e^{-3t}\right)\begin{bmatrix} 0 & 1 & 0 \\ 0 & 0 & 1 \\ -6 & -11 & -6 \end{bmatrix}$$

$$+ \left(\frac{1}{2}e^{-t} - e^{-2t} + \frac{1}{2}e^{-3t}\right)\begin{bmatrix} 0 & 0 & 1 \\ -6 & -11 & -6 \\ 36 & 60 & 25 \end{bmatrix}$$

or

$$e^{At} = \begin{bmatrix} (\ e^{-3t} \ -3e^{-2t} \ +3e^{-t}) & (\frac{3}{2}e^{-3t} \ -4e^{-2t} \ +\frac{5}{2}e^{-t}) & (\frac{1}{2}e^{-3t} \ -e^{-2t} \ +\frac{1}{2}e^{-t}) \\ (-3e^{-3t} \ +6e^{-2t} \ -3e^{-t}) & (-\frac{9}{2}e^{-3t} \ +8e^{-2t} \ -\frac{5}{2}e^{-t}) & (-\frac{3}{2}e^{-3t} \ +2e^{-2t} \ -\frac{1}{2}e^{-t}) \\ (\ 9e^{-3t} \ -12e^{-2t} \ +3e^{-t}) & (\frac{27}{2}e^{-3t} \ -16e^{-2t} \ +\frac{5}{2}e^{-t}) & (\frac{9}{2}e^{-3t} \ -4e^{-2t} \ +\frac{1}{2}e^{-t}) \end{bmatrix}$$

This is the same result, obtained in a more expeditious manner, as in the example in Sec. 8.6 which employed Sylvester's theorem.

8.8 MATRIX FUNCTIONS WITH REPEATED EIGENVALUES

Suppose that the eigenvalues of the $n \times n$ matrix A are not all distinct and that some of them are repeated. Let one of the eigenvalues of A have a multiplicity of m; that is, the particular eigenvalue is repeated m times. In this event one of the equations used to obtain the coefficients of $r(\lambda)$ is repeated m times.

Assume, for example that the eigenvalues of A are $\lambda_1, \lambda_1, \lambda_1, \lambda_4, \cdots,$ λ_n. Then

$$\alpha_0 + \alpha_1\lambda_1 + \alpha_2\lambda_1^2 + \alpha_3\lambda_1^3 + \alpha_4\lambda_1^4 + \cdots + \alpha_n\lambda_1^n = f(\lambda_1)$$

$$\alpha_0 + \alpha_1\lambda_1 + \alpha_2\lambda_1^2 + \alpha_3\lambda_1^3 + \alpha_4\lambda_1^4 + \cdots + \alpha_n\lambda_1^n = f(\lambda_1)$$

$$\alpha_0 + \alpha_1\lambda_1 + \alpha_2\lambda_1^2 + \alpha_3\lambda_1^3 + \alpha_4\lambda_1^4 + \cdots + \alpha_n\lambda_1^n = f(\lambda_1)$$

$$\alpha_0 + \alpha_1\lambda_4 + \alpha_2\lambda_4^2 + \alpha_3\lambda_4^3 + \alpha_4\lambda_4^4 + \cdots + \alpha_n\lambda_4^n = f(\lambda_4)$$

$$\cdots$$

$$\alpha_0 + \alpha_1\lambda_n + \alpha_2\lambda_n^2 + \alpha_3\lambda_n^3 + \alpha_4\lambda_n^4 + \cdots + \alpha_n\lambda_n^n = f(\lambda_n)$$

cannot provide the solution for the α's.

However, if the equation representing the repeated eigenvalue is differentiated $m - 1$ times, a consistent set of equations that possesses a solution is obtained. In the case under discussion, for λ_1 with a multiplicity of $m = 3$, two differentiations with respect to λ yield:

$$\alpha_0 + \alpha_1\lambda_1 + \alpha_2\lambda_1^2 + \alpha_3\lambda_1^3 + \alpha_4\lambda_1^4 + \cdots + \alpha_n\lambda_1^n = f(\lambda_1)$$

$$\alpha_1 + 2\alpha_2\lambda_1 + 3\alpha_3\lambda_1^2 + 4\alpha_4\lambda_1^3 + \cdots + n\alpha_n\lambda_1^{n-1} = f'(\lambda_1)$$

and

$$2\alpha_2 + 6\alpha_3\lambda_1 + 12\alpha_4\lambda_1^2 + \cdots n(n - 1)\lambda_1^{n-2} = f''(\lambda_1)$$

with all other equations involving nonrepeated eigenvalues remaining the same.

Example: It is desired to find e^{At} if

$$A = \begin{bmatrix} -4 & 1 \\ 0 & -4 \end{bmatrix}$$

Here

$$p(\lambda) = (-4 - \lambda)^2 = (\lambda + 4)^2$$

and there are two eigenvalues that are equal:

$$\lambda_1 = \lambda_2 = -4$$

so that

$$f(\lambda_1 = -4) = f(\lambda_2 = -4) = e^{-4t}$$

In this case the two equations that permit the evaluation of the α's in $r(\lambda)$ are:

$$\alpha_0 + \alpha_1\lambda_1 = e^{\lambda_1 t}$$

and its derivative

$$\alpha_1 = -te^{\lambda_1 t}$$

Using $\lambda_1 = \lambda_2 = -4$, these become

$$\alpha_0 - 4\alpha_1 = e^{-4t}$$

$$\alpha_1 = -te^{-4t}$$

It is easy to see that

$$\alpha_1 = -te^{-4t}$$

and

$$\alpha_0 = 4\alpha_1 + e^{-4t} = e^{-4t} - 4te^{-4t}$$

Thus

$$f(A) = e^{-4t} = \alpha_0 I + \alpha_1 A$$

or

$$e^{At} = (e^{-4t} - 4te^{-4t}) \begin{bmatrix} 1 & 0 \\ 0 & 1 \end{bmatrix} - (te^{-4t}) \begin{bmatrix} -4 & 1 \\ 0 & -4 \end{bmatrix}$$

$$e^{At} = \begin{bmatrix} e^{-4t} & -te^{-4t} \\ 0 & e^{-4t} \end{bmatrix}$$

8.9 ANOTHER METHOD

The study of the eigenvalue problem in Chapter 7 showed that if A is $n \times n$, it possesses n eigenvalues and n eigenvectors associated with the n eigenvalues. If the eigenvectors are linearly independent, then an $n \times n$ modal matrix M may be formed from the eigenvectors and the similarity transformation, from Eq. (7.15),

$$S = M^{-1}AM$$

can be used to diagonalize A into the spectral matrix S composed of the eigenvalues of A:

$$S = \begin{bmatrix} \lambda_1 & 0 & 0 & \cdots & 0 \\ 0 & \lambda_2 & 0 & \cdots & 0 \\ 0 & 0 & \lambda_3 & \cdots & 0 \\ & & \cdots & & \\ 0 & 0 & 0 & \cdots & \lambda_n \end{bmatrix} \tag{8.14}$$

In addition, it has been shown in Sec. 7.13 that the matrix A can be reclaimed from its spectral matrix via

$$A = MSM^{-1} \tag{8.15}$$

If $B = MAM^{-1}$, then a representation for B^m can be found by mathematical induction. First, observe that

$$B^2 = (MAM^{-1})(MAM^{-1}) = MA(M^{-1}M)AM^{-1} = MA^2M$$

and then note that

$$B^3 = (MAM^{-1})(MAM^{-1})(MAM^{-1})$$
$$= MA(MM^{-1})A(MM^{-1})AM^{-1} = MA^3M^{-1}$$

Then, by mathematical induction, the representation for B^m becomes

$$B^m = MA^mM^{-1} \tag{8.16}$$

If Eq. (8.16) is inserted into the matrix polynomial $f(B)$,

$$f(B) = \alpha_0 I + \alpha_1 B + \alpha_2 B^2 + \alpha_3 B^3 \cdots$$

the result is

$$f(B) = \alpha_0 I + \alpha_1 MAM^{-1} + \alpha_2 MA^2M^{-1} + \alpha_3 MA^3M^{-1} + \cdots$$

or

$$f(B) = Mf(A)M^{-1} \tag{8.17}$$

In addition, the mth power of the spectral matrix S defined by Eq. (8.14) is:

$$S^m = \begin{bmatrix} \lambda_1^m & 0 & 0 & \cdots & 0 \\ 0 & \lambda_2^m & 0 & \cdots & 0 \\ 0 & 0 & \lambda_3^m & \cdots & 0 \\ & & \cdots & & \\ 0 & 0 & 0 & \cdots & \lambda_n^m \end{bmatrix}$$

and any polynomial $f(S)$ can be written as

$$f(S) = \begin{bmatrix} f(\lambda_1) & 0 & 0 & \cdots & 0 \\ 0 & f(\lambda_2) & 0 & \cdots & 0 \\ 0 & 0 & f(\lambda_3) & \cdots & 0 \\ & & \cdots & & \\ 0 & 0 & 0 & \cdots & f(\lambda_n) \end{bmatrix}$$

Thus by $A = MSM^{-1}$, one may write $f(A)$ as

$$f(A) = f(MSM^{-1}) = Mf(S)M^{-1}$$

or

$$f(A) = Mf(\lambda)M^{-1} \tag{8.18}$$

Equation (8.18) affords what may well be the simplest method for the evaluation of certain matrix functions.

Example: Again suppose that it is desired to determine $f(A) = e^{At}$ if

$$A = \begin{bmatrix} 0 & 1 & 0 \\ 0 & 0 & 1 \\ -6 & -11 & -6 \end{bmatrix}$$

As shown in Section 7.13, the matrix A has three eigenvalues, $\lambda_1 = -1$, $\lambda_2 = -2$, and $\lambda_3 = -3$ and three eigenvectors that form the modal matrix M:

$$M = \begin{bmatrix} -1 & -1/2 & -1/3 \\ 1 & 1 & 1 \\ -1 & -2 & -3 \end{bmatrix}$$

which possesses an inverse:

$$M^{-1} = \begin{bmatrix} -3 & -5/2 & -1/2 \\ 6 & 8 & 2 \\ -3 & -9/2 & -3/2 \end{bmatrix}$$

Application of Eq. (8.18) provides

$$e^{At} = \begin{bmatrix} -1 & -1/2 & -1/3 \\ 1 & 1 & 1 \\ -1 & -2 & -3 \end{bmatrix} \begin{bmatrix} e^{-t} & 0 & 0 \\ 0 & e^{-2t} & 0 \\ 0 & 0 & e^{-3t} \end{bmatrix} \begin{bmatrix} -3 & -5/2 & -1/2 \\ 6 & 8 & 2 \\ -3 & -9/2 & -3/2 \end{bmatrix}$$

$$= \begin{bmatrix} -e^{-t} & -\frac{1}{2}e^{-2t} & -\frac{1}{3}e^{-3t} \\ e^{-t} & e^{-2t} & e^{-3t} \\ -e^{-t} & -2e^{-2t} & -3e^{-3t} \end{bmatrix} \begin{bmatrix} -3 & -5/2 & -1/2 \\ 6 & 8 & 2 \\ -3 & -9/2 & -3/2 \end{bmatrix}$$

or

$$
e^{At} = \begin{bmatrix}
(\ e^{-3t} & -3e^{-2t} & +3e^{-t}) \\
(-s3e^{-3t} & +6e^{-2} & -3e^{-t}) \\
(\ 9e^{-3t} & -12e^{-2t} & +3e^{-t})
\end{bmatrix}
$$

$$
\begin{matrix}
(\ \frac{3}{2}e^{-3t} & -4e^{-2t} & +\frac{5}{2}e^{-t}) & (\ \frac{1}{2}e^{-3t} & -e^{-2t} & +\frac{1}{2}e^{-t}) \\
(-\frac{9}{2}e^{-3t} & +8e^{-2t} & -\frac{5}{2}e^{-t}) & (-\frac{3}{2}e^{-3t} & +2e^{-2t} & -\frac{1}{2}e^{-t}) \\
(\ \frac{27}{2}e^{-3t} & -16e^{-2t} & +\frac{5}{2}e^{-t}) & (\ \frac{9}{2}e^{-3t} & -4e^{-2t} & +\frac{1}{2}e^{-t})
\end{matrix}
$$

This is the same result that was obtained in Sections 8.6 and 8.7.

8.10 THE SOLUTION OF DIFFERENTIAL EQUATIONS

Consider the nth-order ordinary linear differential equation with constant coefficients with dependent variable y and independent variable t:

$$
\frac{d^n y}{dt^n} + \alpha_n \frac{d^{n-1}y}{dt^{n-1}} + \alpha_{n-1}\frac{d^{n-2}y}{dt^{n-2}} + \cdots + \alpha_3 \frac{d^2 y}{dt^2} + \alpha_2 \frac{dy}{dt} + \alpha_1 y = f(t)
$$

and define n new variables $f(t)$, noting that the functional representation is $x = f(t)$:

$$x_1(t) = y(t)$$

$$x_2(t) = \frac{dy}{dt}$$

$$x_3(t) = \frac{d^2 y}{dt^2}$$

$$\cdots$$

$$x_n(t) = \frac{d^{n-1}y}{dt^{n-1}}$$

The derivative of each of these may be taken, using primes to denote the derivatives:

$$x_1'(t) = y'(t)$$
$$x_2'(t) = y''(t)$$
$$x_3'(t) = y'''(t)$$

$$\cdots$$

$$x_n'(t) = y^{(n)}(t)$$

and these may be rearranged to yield a set of n first-order differential equations:

$$x_1'(t) = x_2(t)$$
$$x_2'(t) = x_3(t)$$
$$x_3'(t) = x_4(t)$$
$$\cdots$$
$$x_n'(t) = f(t) - \alpha_1 x_1(t) - \alpha_2 x_2(t) - \cdots$$

which may be put into the matrix form often referred to as the *state variable form*:

$$X' = AX + Bf(t) \tag{8.19}$$

Here the matrix A is defined by:

$$A = \begin{bmatrix} 0 & 1 & 0 & 0 & \cdots & 0 \\ 0 & 0 & 1 & 0 & \cdots & 0 \\ 0 & 0 & 0 & 1 & \cdots & 0 \\ & & \cdots & & & \\ -\alpha_1 & -\alpha_2 & -\alpha_3 & -\alpha_4 & \cdots & -\alpha_n \end{bmatrix} \tag{8.20}$$

and the column vector B is:

$$B = \begin{bmatrix} 0 \\ 0 \\ 0 \\ \cdots \\ 1 \end{bmatrix} \tag{8.21}$$

Notice that the nth-order linear differential equation has been reduced to a set of n simultaneous, linear first-order differential equations. These are easily solved through the use of the Laplace transformation or via a matric integrating vector. The use of the integrating factor is demonstrated in Section 8.11.

Example: The linear third-order differential equation with constant coefficients:

$$\frac{d^3y}{dt^3} + 6\frac{d^2y}{dt^2} + 11\frac{dy}{dt} + 6y = 24$$

can be represented by the set of three simultaneous linear differential equations:

$$
\begin{bmatrix} x_1' \\ x_2' \\ x_3' \end{bmatrix} = \begin{bmatrix} 0 & 1 & 0 \\ 0 & 0 & 1 \\ -6 & -11 & -6 \end{bmatrix} \begin{bmatrix} x_1 \\ x_2 \\ x_3 \end{bmatrix} + \begin{bmatrix} 0 \\ 0 \\ 1 \end{bmatrix} 24
$$

8.11 USE OF THE INTEGRATING FACTOR AND THE SOLUTION METHOD

The solution of a simultaneous set of first-order equations such as the set indicated, in matrix form, by Eq. (8.19) can be obtained through the use of an exponential integrating factor. The procedure is identical to the procedure employed for a single first-order differential equation.

Consider the first-order differential equation

$$
\frac{dy}{dt} + 2y = 12
$$

where $y(0) = 4$. Because it is known that an exponential solution exists, an integrating factor e^{2t} may be selected and the entire differential equation multiplied by this integrating factor. The result is:

$$
e^{2t} \frac{dy}{dt} + 2ye^{2t} = 12e^{2t}
$$

Notice that the left-hand side of this equation is composed entirely of the derivative of ye^{2t}. Thus the differential equation has been adjusted to the form:

$$
\frac{d}{dt}(ye^{2t}) = 12e^{2t}
$$

and an integration yields:

$$
ye^{2t} = 6e^{2t} + \gamma
$$

where γ is an arbitrary constant and where a division throughout by e^{2t} yields the following general solution:

$$
y = \gamma e^{-2t} + 6
$$

This particular solution is obtained by noting that at $t = 0$, $y = 4$. Hence γ is easily evaluated:

$$
4 = \gamma + 6
$$

or $\gamma = -2$. Thus

$$y = 6 - 2e^{-2t}$$

is the solution.

The set of equations is solved by a similar procedure. Begin with a rearrangement of Eq. (8.19):

$$X' - AX = Bf(t)$$

and premultiply by e^{-At} to obtain:

$$e^{-At}(X' - AX) = e^{-At}Bf(t)$$

or

$$\frac{d}{dt}(e^{-At}X) = e^{-At}Bf(t)$$

An integration between 0 and t provides:

$$e^{-At}X(t) - X(0) = \int_0^t e^{-Az}Bf(z)dz$$

so that a premultiplication by $(e^{-At})^{-1}$ yields:

$$X(t) = X = e^{At}X(0) + \int_0^t e^{A(t-z)}Bf(z)dz \qquad (8.22)$$

Example: The solution to the differential equation

$$\frac{d^2y}{dt^2} + 5\frac{dy}{dt} + 6y = 16$$

where $y(0) = 0$ and $dy/dt|_{t=0} = 4$ is required.
The differential equation can be recast in the form of Eq. (8.19):

$$\begin{bmatrix} x_1' \\ x_2' \end{bmatrix} = \begin{bmatrix} 0 & 1 \\ -6 & -5 \end{bmatrix}\begin{bmatrix} x_1 \\ x_2 \end{bmatrix} + \begin{bmatrix} 0 \\ 1 \end{bmatrix}16$$

with an initial condition vector:

$$X(0) = \begin{bmatrix} x_1(0) \\ x_2(0) \end{bmatrix} = \begin{bmatrix} 0 \\ 4 \end{bmatrix}$$

Here

$$A = \begin{bmatrix} 0 & 1 \\ -6 & -5 \end{bmatrix}$$

which has eigenvalues $\lambda_1 = -2$ and $\lambda_2 = -3$.

The matrix A possesses a modal matrix:

$$M = \begin{bmatrix} 1 & 1 \\ -2 & -3 \end{bmatrix}$$

that has an inverse:

$$M^{-1} = \begin{bmatrix} 3 & 1 \\ -2 & -1 \end{bmatrix}$$

so that e^{At} becomes:

$$e^{At} = \begin{bmatrix} 1 & 1 \\ -2 & -3 \end{bmatrix} \begin{bmatrix} e^{-2t} & 0 \\ 0 & e^{-3t} \end{bmatrix} \begin{bmatrix} 3 & 1 \\ -2 & -1 \end{bmatrix}$$

$$= \begin{bmatrix} e^{-2t} & e^{-3t} \\ -2e^{-2t} & -3e^{-3t} \end{bmatrix} \begin{bmatrix} 3 & 1 \\ -2 & -1 \end{bmatrix}$$

or

$$e^{At} = \begin{bmatrix} (3e^{-2t} - 2e^{-3t}) & (e^{-2t} - e^{-3t}) \\ (6e^{-3t} - 6e^{-2t}) & (3e^{-3t} - 2e^{-2t}) \end{bmatrix}$$

Consider the solution for X, as given by Eq. (8.22), to be composed of two parts, one involving the initial condition vector and one involving the integration. The initial condition portion is:

$$e^{At}X(0) = \begin{bmatrix} (3e^{-2t} - 2e^{-3t}) & (e^{-2t} - e^{-3t}) \\ (6e^{-3t} - 6e^{-2t}) & (3e^{-3t} - 2e^{-2t}) \end{bmatrix} \begin{bmatrix} 0 \\ 4 \end{bmatrix}$$

$$= \begin{bmatrix} (4e^{-2t} - 4e^{-3t}) \\ (12e^{-3t} - 8e^{-2t}) \end{bmatrix}$$

Notice that the portion involving the integral requires the evaluation of the matrix product $Bf(t)$, which is seen to be:

$$Bf(t) = \begin{bmatrix} 0 \\ 16 \end{bmatrix}$$

Thus the term within the integral to be evaluated is:

$$\begin{bmatrix} (3e^{-2(t-z)} - 2e^{-3(t-z)}) & (e^{-2(t-z)} - e^{-3(t-z)}) \\ (6e^{-3(t-z)} - 6e^{-2(t-z)}) & (3e^{-3(t-z)} - 2e^{-2(t-z)}) \end{bmatrix} \begin{bmatrix} 0 \\ 16 \end{bmatrix}$$

$$= \begin{bmatrix} (16e^{-2(t-z)} - 16e^{-3(t-z)}) \\ (48e^{-3(t-z)} - 32e^{-2(t-z)}) \end{bmatrix}$$

The integration may then be performed:

$$\int_0^t e^{A(t-z)} Bf(z)dz = \int_0^t \begin{bmatrix} (16e^{-2(t-z)} - 16e^{-3(t-z)}) \\ (48e^{-3(t-z)} - 32e^{-2(t-z)}) \end{bmatrix} dz$$

$$= \begin{bmatrix} 8e^{-2(t-z)} - \frac{16}{3}e^{-3(t-z)} \\ 16e^{-3(t-z)} - 16e^{-2(t-z)} \end{bmatrix}_0^t$$

so that the solution portion of Eq. (8.22) involving the integral becomes

$$\int_0^t e^{A(t-z)} Bf(z)dz = \begin{bmatrix} (\frac{16}{3}e^{-3t} - 8e^{-2t} + \frac{8}{3}) \\ (16e^{-2t} - 16e^{-3t}) \end{bmatrix}$$

When the two solution components are put together, the result is:

$$X = \begin{bmatrix} x_1(t) \\ x_2(t) \end{bmatrix} = \begin{bmatrix} (\frac{4}{3}e^{-3t} - 4e^{-2t} + \frac{8}{3}) \\ (8e^{-2t} - 4e^{-3t}) \end{bmatrix}$$

from which

$$y(t) = x_1(t) = \frac{4}{3}e^{-3t} - 4e^{-2t} + \frac{8}{3}$$

and the reader may wish to verify that this is the solution to the second-order differential equation originally posed. Moreover, it can be noted that $x_2(t)$ is the derivative of $x_1(t)$. This fact can be used to further demonstrate the validity of the solution.

8.12 SOLUTION OF SIMULTANEOUS DIFFERENTIAL EQUATIONS

The method of solution demonstrated in the previous section is even more useful when applied to the solution of what is called the *state variable* or *state space problem*. This is an application of the method in solving n simultaneous linear first-order differential equations in n unknown state variables.

Example: Consider the system of first-order differential equations with $x_1(0) = 1$ and $x_2(0) = -2$:

$$\frac{dx_1}{dt} + 5x_1 + 3x_2 = 0$$

$$\frac{dx_2}{dt} - 3x_1 - x_2 = 0$$

and set up the state variable form of Eq. (8.19):

$$X = \begin{bmatrix} x_1'(t) \\ x_2'(t) \end{bmatrix} = \begin{bmatrix} -5 & -3 \\ 3 & 1 \end{bmatrix} \begin{bmatrix} x_1(t) \\ x_2(t) \end{bmatrix}$$

Observe that because the system of equations is homogeneous, there is no need to consider the vector B.

The characteristic polynomial of the matrix A:

$$A = \begin{bmatrix} -5 & -3 \\ 3 & 1 \end{bmatrix}$$

is easily determined:

$$p(\lambda) = \lambda^2 + 4\lambda + 4 = 0$$

and it is seen that there are two identical eigenvalues:

$$\lambda_1 = \lambda_2 = -2$$

The method of Sec. 8.8 must be used to evaluate e^{At}. The two equations needed to determine the two constants α_0 and α_1 in

$$e^{At} = \alpha_0 I + \alpha_1 A$$

are

$$\alpha_0 - 2\alpha_1 = e^{-2t}$$

and

$$\alpha_1 = -te^{-2t}$$

Solution of these yields

$$\alpha_1 = te^{-2t}$$

and

$$\alpha_0 = e^{-2t} + 2te^{-2t}$$

so that

$$e^{At} = (e^{-2t} + 2te^{-2t})\begin{bmatrix} 1 & 0 \\ 0 & 1 \end{bmatrix} + te^{-2t}\begin{bmatrix} -5 & -3 \\ 3 & 1 \end{bmatrix}$$

or

$$e^{At} = \begin{bmatrix} (e^{-2t} - 3te^{-2t}) & (-3te^{-2t}) \\ (3te^{-2t}) & (e^{-2t} + 3te^{-2t}) \end{bmatrix}$$

Now with the initial condition vector:

$$X(0) = \begin{bmatrix} 1 \\ -2 \end{bmatrix}$$

it is easy to get both $x_1(t)$ and $x_2(t)$ at the same time via a simple application of Eq. (8.22):

$$X = \begin{bmatrix} x_1(t) \\ x_2(t) \end{bmatrix} = \begin{bmatrix} (e^{-2t} - 3te^{-2t}) & (-3te^{-2t}) \\ (3te^{-2t}) & (e^{-2t} + 3te^{-2t}) \end{bmatrix} \begin{bmatrix} 1 \\ -2 \end{bmatrix}$$

or

$$X = \begin{bmatrix} x_1(t) \\ x_2(t) \end{bmatrix} = \begin{bmatrix} (e^{-2t} + 3te^{-2t}) \\ (-2e^{-2t} - 3te^{-2t}) \end{bmatrix}$$

The reader may wish to verify the foregoing result and to observe that in this pair of simultaneous linear and homogeneous first-order differential equations, $x_2(t)$ bears no resemblence whatsoever to the derivative of $x_1(t)$.

8.13 EXERCISES

Exercises 8.1 through 8.6 pertain to the matrix A:

$$A = \begin{bmatrix} -1 & 4 & -2 \\ -3 & 4 & 0 \\ -3 & 1 & 3 \end{bmatrix}$$

8.1 Use the Cayley–Hamilton theorem to evaluate

$$f(A) = A^5 - 4A^4 + 3A^3 + A^2 + 4A - I$$

8.2 Use Sylvester's theorem to evaluate $f(A)$ in Exercise 8.1.

8.3 Use Sylvester's theorem to find e^{At}.

8.4 Use the Cayley–Hamilton theorem to find A^{-1}.

8.5 Find a matrix M that will diagonalize A.

8.6 Confirm the result of Exercise 8.3 by another method.

Exercise 8.7 through 8.24 pertain to the matrices A and B:

$$A = \begin{bmatrix} -3 & 1 \\ 0 & -2 \end{bmatrix} \qquad B = \begin{bmatrix} -3/2 & 1/2 \\ 0 & -1 \end{bmatrix}$$

8.7 Find e^A.

8.8 Find e^{-A}.

8.9 Evaluate $(e^A)(-e^A)$.

8.10 Evaluate $e^A e^{-A}$.

8.11 Find e^B.

8.12 Does $e^B e^B = e^A$?

8.13 Find $\cosh A$.

8.14 Does $\cosh A = \frac{1}{2}(e^A + e^{-A})$?

8.15 Find $\sinh A$.

8.16 Does $\sinh A = \frac{1}{2}(e^A - e^{-A})$?

8.17 Find $\tanh A$.

8.18 Find $\cos A$.

8.19 Find $\sin A$.

8.20 Find $\cos B$.

8.21 Find $\sin B$.

8.22 Find $\cos (A - B)$. Does $\cos (A - B)$ satisfy the following identity:

$$\cos (A - B) = \cos A \cos B - \sin A \sin B$$

8.23 Find $\sin^2 A + \cos^2 A$. Does $\sin^2 A + \cos^2 A$ satisfy the following identity:

$$\sin^2 A + \cos^2 A = I$$

8.24 What conclusion should be drawn from the results of Exercises 8.22 and 8.23?

NINE

EXAMPLES

9.1 INTRODUCTION

This chapter is concerned with the provision of some interdisciplinary examples of typical analyses that are greatly simplified by the matrix methods developed in the previous chapters.

9.2 STEADY-STATE DC ELECTRICAL NODE-TO-DATUM AND MESH ANALYSIS

9.2.1 Preliminaries

The matrix approach to dc network node-to-datum and mesh analysis is based on several preliminary ideas:

1. The resistances (the R's) are lumped elements.
2. Lumped networks may be formed by connecting lumped elements.
3. Lumped elements possess two terminals.
4. Two (or more) terminals, when connected together, form a node.
5. Single lumped elements form branches.
6. Ideal voltage sources when connected with a lumped resistor in series form a branch.
7. Ideal current sources when connected with a lumped resistor in parallel form a branch.

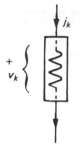

Figure 9.1 The general resistive branch without sources.

Use of the techniques outlined in this section requires a resistor in series with a voltage source and in parallel with a current source.

A single resistive element, which may be considered as a general branch (the kth branch), is displayed in Fig. 9.1. Observe that this element contains no sources. The branch current is designated by a j (i's are reserved for mesh or loop currents) and the branch voltage, which is often called a voltage drop, is designated by a v (e's are reserved for node-to-datum voltages). The voltage v is oriented with its high or plus side at the entry point of the branch current. It is useful to think of the branch current as proceeding in the direction of the branch voltage drop.

The kth branch with sources is shown in Fig. 9.2. Notice that to be consistent with the convention adopted for the branch without sources (Fig. 9.1), the overall branch current flows in the direction of the voltage drop in the voltage source and in the direction of the source current. Also notice that a resistive element must be contained in the branch and that the branch may possess both voltage and current sources. There are, of course, two possibilities (Figs. 9.2a and b), depending on whether the current source is in parallel with just the resistive element or with both

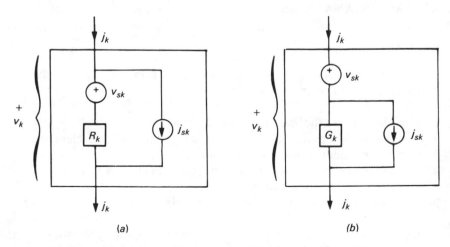

(a)   (b)

Figure 9.2 Two representations of the general resistive branch with sources.

the resistive element and the voltage source. This really amounts to a consideration of whether the current source is in parallel or in series with the voltage source.

For Fig. 9.2a where the voltage and current sources are designated respectively as v_{sk} and j_{sk}, an application of the Kirchhoff current law shows that the current through the voltage source is equal to

$$v_k = v_{sk} + R_k(j_k - j_{sk})$$

or

$$v_k = R_k j_k + v_{sk} - R_k j_{sk}$$

In matrix form this becomes

$$V = RJ + V_s - RJ_s \tag{9.1}$$

where for b branches V, V_s, and J_s are $b \times 1$ column vectors and where R is a diagonal $b \times b$ matrix containing the b resistance values on its principal diagonal. Observe that each branch must contain a resistor but not all branches contain sources.

For Fig. 9.2b a similar procedure can be employed, but with the Kirchoff voltage law used first, to yield the following:

$$j_k = j_{sk} + G_k(v_k - v_{sk})$$

or

$$j_k = G_k v_k + j_{sk} - G_k v_{sk}$$

and in matrix form:

$$J = GV + J_s - GV_s \tag{9.2}$$

Here the additional matrix is the $b \times b$ conduction matrix G, which is diagonal and which has no element on the principal diagonal equal to zero. Each value of G is equal to the reciprocal of a particular $R(G = 1/R)$.

9.2.2 The Concept of a Graph and Oriented Graphs

A linear network graph is a collection of branches and nodes that is arranged to represent the geometry of a network exactly. The graph does not depend on the actual elements of the network or their magnitude. Its purpose is to assist the determination of network properties such as voltages and currents based on structure, geometry, or interconnection. When each branch of a graph carries an arrow to indicate its orientation (its branch current direction), the graph is said to be an *oriented graph*.

Orientation of branches is partially discretionary. Branches that contain sources must be oriented in a manner consistent with the convention required by the general branch with sources, as shown in Fig. 9.2.

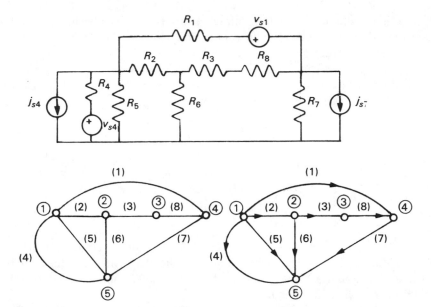

Figure 9.3 (*a*) A network containing eight branches with two current sources and two voltage sources; (*b*) its graph showing branch and node identifiers; and (*c*) an oriented graph.

Figure 9.3*a* shows a network containing eight resistors, two voltage sources, and two current sources. A graph containing eight branches with branch designators in parentheses is shown in Fig. 9.3*b*. Here the branch designators correspond with the subscripts contained on the eight resistors. The nodes are designated with numerals enclosed in circles. The oriented graph is displayed in Fig. 9.3*c*. The orientations are discretionary for branches 2, 3, 5, 6, and 8, but the orientations are in accordance with Fig. 9.2 for branches 1, 4, and 7.

From the oriented graph containing the eight branches, it is easy to see that the vectors V_s and J_s are the $b \times 1$ ($b = 8$) column vectors:

$$V_s = \begin{bmatrix} v_{s1} \\ 0 \\ 0 \\ v_{s4} \\ 0 \\ 0 \\ 0 \\ 0 \end{bmatrix} \qquad J_s = \begin{bmatrix} 0 \\ 0 \\ 0 \\ j_{s4} \\ 0 \\ 0 \\ j_{s7} \\ 0 \end{bmatrix}$$

The $b \times b$ ($b = 8$) R matrix is:

$$R = \begin{bmatrix} R_1 & 0 & 0 & 0 & 0 & 0 & 0 & 0 \\ 0 & R_2 & 0 & 0 & 0 & 0 & 0 & 0 \\ 0 & 0 & R_3 & 0 & 0 & 0 & 0 & 0 \\ 0 & 0 & 0 & R_4 & 0 & 0 & 0 & 0 \\ 0 & 0 & 0 & 0 & R_5 & 0 & 0 & 0 \\ 0 & 0 & 0 & 0 & 0 & R_6 & 0 & 0 \\ 0 & 0 & 0 & 0 & 0 & 0 & R_7 & 0 \\ 0 & 0 & 0 & 0 & 0 & 0 & 0 & R_8 \end{bmatrix}$$

and with each $G = 1/R$, the $b \times b$ ($b = 8$) G matrix is:

$$G = \begin{bmatrix} G_1 & 0 & 0 & 0 & 0 & 0 & 0 & 0 \\ 0 & G_2 & 0 & 0 & 0 & 0 & 0 & 0 \\ 0 & 0 & G_3 & 0 & 0 & 0 & 0 & 0 \\ 0 & 0 & 0 & G_4 & 0 & 0 & 0 & 0 \\ 0 & 0 & 0 & 0 & G_5 & 0 & 0 & 0 \\ 0 & 0 & 0 & 0 & 0 & G_6 & 0 & 0 \\ 0 & 0 & 0 & 0 & 0 & 0 & G_7 & 0 \\ 0 & 0 & 0 & 0 & 0 & 0 & 0 & G_8 \end{bmatrix}$$

Moreover, the sought-after branch currents and branch voltages can be placed into a pair of $b \times 1$ ($b = 8$) column vectors, known respectively as the *branch voltage* and *branch current vectors*

$$V = \begin{bmatrix} v_1 \\ v_2 \\ v_3 \\ v_4 \\ v_5 \\ v_6 \\ v_7 \\ v_8 \end{bmatrix} \qquad J = \begin{bmatrix} j_1 \\ j_2 \\ j_3 \\ j_4 \\ j_5 \\ j_6 \\ j_7 \\ j_8 \end{bmatrix}$$

9.2.3 Node-to-Datum Analysis

The oriented graph in Fig 9.3c contains eight branches ($b = 8$) and a total of five nodes ($n_t = 5$). The augmented node-branch incidence matrix is an $n_t \times b$ matrix A^a having elements a_{ij} where

$$
a_{ij} = \begin{cases} +1 \text{ if branch } j \text{ leaves node } i \\ -1 \text{ if branch } j \text{ enters node } i \\ 0 \text{ if branch } j \text{ does not touch node } i \end{cases}
$$

For the network of Fig. 9.3a with its oriented graph of Fig 9.3c,

$$
A^a = n_t \quad
\begin{array}{c}
1 \\ 2 \\ 3 \\ 4 \\ 5
\end{array}
\begin{bmatrix}
1 & 1 & 0 & 1 & 1 & 0 & 0 & 0 \\
0 & -1 & 1 & 0 & 0 & 1 & 0 & 0 \\
0 & 0 & -1 & 0 & 0 & 0 & 0 & 1 \\
-1 & 0 & 0 & 0 & 0 & 0 & 1 & -1 \\
0 & 0 & 0 & -1 & -1 & -1 & -1 & 0
\end{bmatrix}
$$

with the column headers $1\ 2\ 3\ 4\ 5\ 6\ 7\ 8$ labeled b.

Observe that every column of A^a contains a single 1 and a single -1, no more, no less. All other elements in each column are zeroes.

The node-branch incidence matrix A derives from the augmented node-branch incidence matrix. One of the nodes is selected as the datum or ground node. If the row of A^a corresponding to the datum or ground node is deleted, the node-branch incidence matrix results. Here, with the node 5 selected as the datum node,

$$
A = \begin{bmatrix}
1 & 1 & 0 & 1 & 1 & 0 & 0 & 0 \\
0 & -1 & 1 & 0 & 0 & 1 & 0 & 0 \\
0 & 0 & -1 & 0 & 0 & 0 & 0 & 1 \\
-1 & 0 & 0 & 0 & 0 & 0 & 1 & -1
\end{bmatrix}
$$

Now consider that the sought-after branch current matrix can be pre-multiplied by the node-branch incidence matrix:

$$AJ = \begin{bmatrix} 1 & 1 & 0 & 1 & 1 & 0 & 0 & 0 \\ 0 & -1 & 1 & 0 & 0 & 1 & 0 & 0 \\ 0 & 0 & -1 & 0 & 0 & 0 & 0 & 1 \\ -1 & 0 & 0 & 0 & 0 & 0 & 1 & -1 \end{bmatrix} \begin{bmatrix} j_1 \\ j_2 \\ j_3 \\ j_4 \\ j_5 \\ j_6 \\ j_7 \\ j_8 \end{bmatrix}$$

$$= \begin{bmatrix} (j_1 + j_2 + j_4 + j_5) \\ (-j_2 + j_3 + j_6) \\ (-j_3 + j_8) \\ (-j_1 + j_7 - j_8) \end{bmatrix}$$

A comparison of this result with an application of the Kirchhoff current law at nodes 1 through 4 in Fig. 9.3c can be made. The comparison shows that the algebraic sum of all branch currents entering or leaving each of nodes 1 through 4 is equal to zero. Thus

$$\begin{bmatrix} (j_1 + j_2 + j_4 + j_5) \\ (-j_2 + j_3 + j_6) \\ (-j_3 + j_8) \\ (-j_1 + j_7 - j_8) \end{bmatrix} = \begin{bmatrix} 0 \\ 0 \\ 0 \\ 0 \end{bmatrix}$$

and this is an expression of the Kirchhoff current law at each of the nodes for this specific network. It holds for any network, and

$$AJ = 0 \qquad (9.3)$$

is an expression of the Kirchhoff current law.

If the node-to-datum voltages are defined by an $n \times 1$ column vector E, where n is one less than the total number of nodes ($n = n_t - 1$), then here where $n = 4$

$$E = \begin{bmatrix} e_1 \\ e_2 \\ e_3 \\ e_4 \end{bmatrix}$$

Each of the branch voltages can be written in terms of one or more of the node-to-datum voltages:

$$v_1 = e_1 - e_4$$

$$v_2 = e_1 - e_2$$

$$v_3 = e_2 - e_3$$

$$v_4 = e_1$$

$$v_5 = e_1$$

$$v_6 = e_2$$

$$v_7 = e_4$$

$$v_8 = e_3 - e_4$$

A matrix formulation of these equations is $CE = V$, where C must be $b \times n$. In this case with $b = 8$ and $n = 4$,

$$
\begin{bmatrix}
1 & 0 & 0 & -1 \\
1 & -1 & 0 & 0 \\
0 & 1 & -1 & 0 \\
1 & 0 & 0 & 0 \\
1 & 0 & 0 & 0 \\
0 & 1 & 0 & 0 \\
0 & 0 & 0 & 1 \\
0 & 0 & 1 & -1
\end{bmatrix}
\begin{bmatrix}
e_1 \\
e_2 \\
e_3 \\
e_4
\end{bmatrix}
=
\begin{bmatrix}
v_1 \\
v_2 \\
v_3 \\
v_4 \\
v_5 \\
v_6 \\
v_7 \\
v_8
\end{bmatrix}
$$

and it is interesting to observe that $C = A^{\mathrm{T}}$. Thus

$$V = A^{\mathrm{T}}E \tag{9.4}$$

and all the equations needed to conduct a node-to-datum analysis are available.

Begin with Eq. (9.2):

$$J = GV + J_s - GV_s$$

and premultiply every term by A:

$$AJ = AGV + AJ_s - AGV_s$$

But in accordance with Eq. (9.3), $AJ = 0$ so that

$$AGV + AJ_s - AGV_s = 0$$

Equation (9.4) may then be used to show that

$$AGA^{T}E = AGV_s - AJ_s$$

This leads to the node-to-datum equations:

$$Y_nE = I_s \tag{9.5}$$

where Y_n is the node admittance matrix:

$$Y_n = AGA^{T} \tag{9.6a}$$

and I_s is the source current vector:

$$I_s = AGV_s - AJ_s \tag{9.6b}$$

9.2.4 Procedure for Node-to-Datum Analysis

The procedure node-to-datum analysis follows a logical sequence:

1. Construct an oriented graph for the network. Be sure that all branches, nodes, and datum node are clearly identified and that the branch orientations are in accordance with Figs. 9.1 and 9.2.
2. Write the node-branch incidence matrix A, the current and voltage source vectors J_s and V_s, and the conductance matrix G.
3. Obtain $Y_n = AGA^{T}$ and $I_s = AGV_s - AJ_s$ and formulate the node-to-datum equations $Y_nE = I_s$.
4. By any computationally efficient procedure, obtain E. In the following example, E is determined through the use of the Cholesky reduction and the reader is invited to compare the labor involved with the determination of E by a matrix inversion of Y_n.
5. Determine the branch voltages using Eq. (9.4), $V = A^{T}E$.
6. Determine the branch currents from Eq. (9.2), $J = GV + J_s - GV_s$.

Example: It is required to use node-to-datum analysis to determine all branch currents and branch voltages in the network of Fig. 9.4a.

The oriented graph of the network is displayed in Fig. 9.4b. Observe that there are seven branches ($b = 7$) and a total of four nodes ($n_t = 4$). The branches can be identified by numerals within parentheses; the nodes by numerals within circles. Branch 1 contains both a current and a voltage source, and the orientation has been selected to correspond with the voltage source. This orientation is opposite to the positive sense of the current source so that, in the current source vector, the element representing the source current, j_{s1}, is negative.

Three matrices may be set down immediately. These are the voltage and current source vectors:

$$V_s = \begin{bmatrix} 12 \\ 0 \\ 16 \\ 0 \\ 0 \\ 0 \\ 24 \end{bmatrix} \qquad J_s = \begin{bmatrix} -2 \\ 0 \\ 4 \\ 16 \\ 0 \\ 0 \\ 0 \end{bmatrix}$$

and the conductance matrix:

$$G = \begin{bmatrix} 1.000 & 0 & 0 & 0 & 0 & 0 & 0 \\ 0 & 0.125 & 0 & 0 & 0 & 0 & 0 \\ 0 & 0 & 1.000 & 0 & 0 & 0 & 0 \\ 0 & 0 & 0 & 0.250 & 0 & 0 & 0 \\ 0 & 0 & 0 & 0 & 0.500 & 0 & 0 \\ 0 & 0 & 0 & 0 & 0 & 1.000 & 0 \\ 0 & 0 & 0 & 0 & 0 & 0 & 0.250 \end{bmatrix}$$

The node-branch incidence matrix, with node 4 selected as the datum node, is:

$$A = \begin{bmatrix} 1 & 0 & 0 & 0 & 1 & 0 & -1 \\ 0 & 1 & 1 & 0 & -1 & 1 & 0 \\ 0 & 0 & 0 & -1 & 0 & -1 & 1 \end{bmatrix}$$

The preliminaries have been concluded, and the next step is to write the node-to-datum equations, $Y_n E = I_s$. Here the node admittance matrix is $Y_n = AGA^\mathrm{T}$. First, obtain AG. The user may verify that the product AG is equal to the following:

$$AG = \begin{bmatrix} 1.000 & 0 & 0 & 0 & 0.500 & 0 & -0.250 \\ 0 & 0.125 & 1.000 & 0 & -0.500 & 1.000 & 0 \\ 0 & 0 & 0 & -0.250 & 0 & -1.000 & 0.250 \end{bmatrix}$$

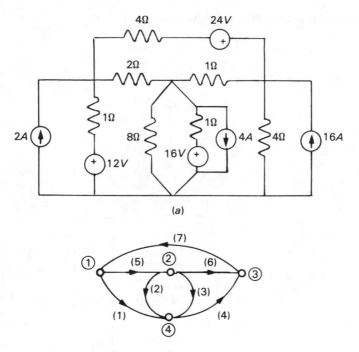

(a)

(b)

Figure 9.4 (*a*) Network to be used to illustrate the procedure for a node-to-datum analysis; (*b*) the oriented graph of the network.

Then with

$$A^T = \begin{bmatrix} 1 & 0 & 0 \\ 0 & 1 & 0 \\ 0 & 1 & 0 \\ 0 & 0 & -1 \\ 1 & -1 & 0 \\ 0 & 1 & -1 \\ -1 & 0 & 1 \end{bmatrix}$$

it is easy to see that $Y_n = AGA^T$ is

$$Y_n = \begin{bmatrix} 1.750 & -0.500 & -0.250 \\ -0.500 & 2.625 & -1.000 \\ -0.250 & -1.000 & 1.500 \end{bmatrix}$$

The current source vector is given by Eq. (9.6b)

$$I_s = AGV_s - AJ_s$$

A premultiplication of V_s by AG yields

$$AGV_s = \begin{bmatrix} 6 \\ 16 \\ 6 \end{bmatrix}$$

and

$$AJ_s = \begin{bmatrix} 1 & 0 & 0 & 0 & 1 & 0 & -1 \\ 0 & 1 & 1 & 0 & -1 & 1 & 0 \\ 0 & 0 & 0 & -1 & 0 & -1 & 1 \end{bmatrix} \begin{bmatrix} -2 \\ 0 \\ 4 \\ 16 \\ 0 \\ 0 \\ 0 \end{bmatrix} = \begin{bmatrix} -2 \\ 4 \\ -16 \end{bmatrix}$$

so that

$$I_s = \begin{bmatrix} 6 \\ 16 \\ 6 \end{bmatrix} - \begin{bmatrix} -2 \\ 4 \\ -16 \end{bmatrix} = \begin{bmatrix} 8 \\ 12 \\ 24 \end{bmatrix}$$

Thus the node-to-datum equations, $Y_n E = I_s$, are:

$$\begin{bmatrix} 1.750 & -0.500 & -0.250 \\ -0.500 & 2.625 & -1.000 \\ -0.250 & -1.000 & 1.500 \end{bmatrix} \begin{bmatrix} e_1 \\ e_2 \\ e_3 \end{bmatrix} = \begin{bmatrix} 8 \\ 12 \\ 24 \end{bmatrix}$$

and the node-to-datum voltage vector E can be obtained easily by a variety of methods. The node admittance matrix (the coefficient matrix for the e's) can be factored using the Cholesky reduction because it is both symmetric and positive definite. The procedure of Section 5.15 using Eqs. (5.19) provides (and a little detail here will most certainly yield a pedagogical benefit)

$$l_{11} = \sqrt{1.750} = 1.3229$$

$$l_{21} = \frac{1}{1.3229}(-0.500) = -0.3780$$

$$l_{22} = \sqrt{2.625 - (-0.3780)^2} = \sqrt{2.4821} = 1.5755$$

$$l_{31} = \frac{1}{1.3229}(-0.250) = -0.1890$$

$$l_{32} = \frac{1}{1.5755}[-1 - (-0.3780)(-0.1890)] = -0.6801$$

and

$$l_{33} = \sqrt{1.500 - (-0.1890)^2 - (-0.6801)^2} = \sqrt{1.0018} = 1.0009$$

Thus $Y_n = LL^T$:

$$Y_n = \begin{bmatrix} 1.3229 & 0 & 0 \\ -0.3780 & 1.5755 & 0 \\ -0.1890 & -0.6801 & 1.0009 \end{bmatrix} \begin{bmatrix} 1.3229 & -0.3780 & -0.1890 \\ 0 & 1.5755 & -0.6801 \\ 0 & 0 & 1.0009 \end{bmatrix}$$

which the reader may wish to verify.

Then by the back substitution algorithm of Eq. (5.20), as given in Section 5.16, with

$$U = \begin{bmatrix} 1.3229 & -0.3780 & -0.1890 \\ 0 & 1.5755 & -0.6801 \\ 0 & 0 & 1.0009 \end{bmatrix}$$

and

$$B = \begin{bmatrix} 8 \\ 12 \\ 24 \end{bmatrix}$$

in $LY = B$, the elements of Y are:

$$y_1 = \frac{1}{1.3229}(8) = 6.0474$$

$$y_2 = \frac{1}{1.5755}[12 - (-0.3780)(6.0474)] = 9.0675$$

and

$$y_3 = \frac{1}{1.0009}[22 - (-0.6801)(9.0675) - (-0.1890)(6.0474)] = 29.2830$$

Then via $UE = L^TE = Y$, the elements of E are found by the algorithm of Eq. (5.21):

$$e_3 = \frac{29.2830}{1.0009} = 29.2567$$

$$e_2 = \frac{1}{1.5755}[9.0675 - (-0.6801)(29.2567)] = 18.3842$$

and

$$e = \frac{1}{1.3229}[6.0474 - (-0.3780)(18.3842) - (-0.1890)(29.2567)]$$

$$= 14.0036$$

This makes the node-to-datum voltage vector (in volts):

$$E = \begin{bmatrix} 14.0036 \\ 18.3842 \\ 29.2567 \end{bmatrix}$$

The reader may wish to try another method such as matrix inversion for the solution of the system of node equations. The amount of work required as compared to the Cholesky reduction can then be noted.

The branch voltages may now be obtained using Eq. (9.4), $V = A^TE$ (in volts):

$$V = \begin{bmatrix} 1 & 0 & 0 \\ 0 & 1 & 0 \\ 0 & 1 & 0 \\ 0 & 0 & -1 \\ 1 & -1 & 0 \\ 0 & 1 & -1 \\ -1 & 0 & 1 \end{bmatrix} \begin{bmatrix} 14.0036 \\ 18.3842 \\ 29.2567 \end{bmatrix} = \begin{bmatrix} 14.0036 \\ 18.3842 \\ 18.3842 \\ -29.2567 \\ -4.3806 \\ -10.8725 \\ 15.2531 \end{bmatrix}$$

and then an application of Eq. (9.2), $J = GV + J_s - GV_s$, provides the branch currents. First, for GV:

$$
GV = \begin{bmatrix}
1.000 & 0 & 0 & 0 & 0 & 0 & 0 \\
0 & 0.125 & 0 & 0 & 0 & 0 & 0 \\
0 & 0 & 1.000 & 0 & 0 & 0 & 0 \\
0 & 0 & 0 & 0.250 & 0 & 0 & 0 \\
0 & 0 & 0 & 0 & 0.500 & 0 & 0 \\
0 & 0 & 0 & 0 & 0 & 1.000 & 0 \\
0 & 0 & 0 & 0 & 0 & 0 & 0.250
\end{bmatrix} \cdot
$$

$$
\begin{bmatrix}
14.0036 \\
18.3842 \\
18.3842 \\
-29.2567 \\
-4.3806 \\
-10.8725 \\
15.2531
\end{bmatrix}
=
\begin{bmatrix}
14.0036 \\
2.2980 \\
18.3842 \\
-7.3142 \\
-2.1903 \\
-10.8725 \\
3.8133
\end{bmatrix}
$$

and then with GV_s easily computed (in amperes):

$$
J =
\begin{bmatrix}
14.0036 \\
2.2980 \\
18.3842 \\
-7.3142 \\
-2.1903 \\
-10.8725 \\
3.8133
\end{bmatrix}
+
\begin{bmatrix}
-2 \\
0 \\
4 \\
16 \\
0 \\
0 \\
0
\end{bmatrix}
-
\begin{bmatrix}
12 \\
0 \\
16 \\
0 \\
0 \\
0 \\
6
\end{bmatrix}
$$

or

$$
J =
\begin{bmatrix}
0.0036 \\
2.2980 \\
6.3842 \\
8.6858 \\
-2.1903 \\
-10.8725 \\
-2.1867
\end{bmatrix}
$$

9.2.5 Mesh Analysis

In a graph a path is formed by beginning at a particular node, traversing one or more branches in sequence, and ending at another node without touching any node more than once. A loop is a closed path that begins and ends at the same node. In Fig. 9.4b the sequence of branches -5, 6, 4 provides a path from node -1 to node -4 as does the sequence of branches -7, 6, 2. The sequences of branches -5, 2, 1 and -5, 6, 4, 1 both constitute loops, because they both describe paths from node -1 to node -1. Observe that the sequence of branches -5, 3, 2, 6, 4 does not constitute a path, because node -2 is touched more than once, and the sequence -5, 6, 4, 2, 3, 1 does not form a loop for the same reason.

A mesh is defined as a specific loop that contains no interior branches or loops. In Fig. 9.4b there are four meshes, all of which are loops. These are formed by branches: 5, 2, 1; 2, 3; 6, 4, 3; and 7, 6, 5. There are many loops that are not meshes. Two of them are formed by branches -7, 6, 3, 1 and -6, 4, 1, 5. Both these loops contain interior branches and loops. One may note that a mesh is always a loop (a specific kind of loop) but that not all loops are meshes.

Mesh analysis of a resistive network begins with the formulation of the $b \times 1$ voltage and current source vectors and the $b \times b$ resistance matrix, which is diagonal and contains the resistance of the branches on

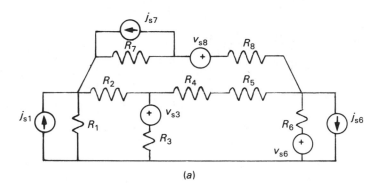

(a)

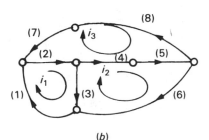

(b)

Figure 9.5 (a) Network; (b) oriented graph showing three meshes with clockwise mesh currents designated by i_1, i_2, and i_3.

the principal diagonal. The actual analysis seeks the $m \times 1$ mesh current vector I, where m is the number of interior meshes. The $m \times 1$ mesh current vector I is related to the $b \times 1$ branch current vector J by the mesh-branch incidence matrix, where the elements m_{ij} are selected in accordance with

$$
m_{ij} = \begin{cases}
+1 \text{ if branch } j \text{ is in mesh } i \text{ and} \\
\quad \text{the branch current is in the direction} \\
\quad \text{of the mesh current} \\
\\
-1 \text{ if branch } j \text{ is in mesh } i \text{ and} \\
\quad \text{the branch current is opposed to the} \\
\quad \text{direction of the mesh current} \\
\\
0 \text{ if branch } j \text{ does not touch mesh } i
\end{cases}
$$

Consider the simple network shown with an oriented graph in Fig. 9.5. Here in Fig. 9.5b the orientations of the branches containing the sources are in accordance with convention adopted in Section 9.2.1. Notice that in Fig. 9.5b three meshes are apparent and that each of them possesses a clockwise mesh current (i_1, i_2, and i_3). Here V_s and J_s are:

$$
V_s = \begin{bmatrix} 0 \\ 0 \\ V_{s3} \\ 0 \\ 0 \\ V_{s6} \\ 0 \\ V_{s8} \end{bmatrix}
\qquad
J_s = \begin{bmatrix} j_{s1} \\ 0 \\ 0 \\ 0 \\ 0 \\ j_{s6} \\ j_{s7} \\ 0 \end{bmatrix}
$$

and the resistance matrix is:

$$
R = \begin{bmatrix}
R_1 & 0 & 0 & 0 & 0 & 0 & 0 & 0 \\
0 & R_2 & 0 & 0 & 0 & 0 & 0 & 0 \\
0 & 0 & R_3 & 0 & 0 & 0 & 0 & 0 \\
0 & 0 & 0 & R_4 & 0 & 0 & 0 & 0 \\
0 & 0 & 0 & 0 & R_5 & 0 & 0 & 0 \\
0 & 0 & 0 & 0 & 0 & R_6 & 0 & 0 \\
0 & 0 & 0 & 0 & 0 & 0 & R_7 & 0 \\
0 & 0 & 0 & 0 & 0 & 0 & 0 & R_8
\end{bmatrix}
$$

The mesh-branch incidence matrix is $m \times b$ (3×8):

$$M = \begin{bmatrix} 1 & 1 & 1 & 0 & 0 & 0 & 0 & 0 \\ 0 & 0 & -1 & 1 & 1 & 1 & 0 & 0 \\ 0 & -1 & 0 & -1 & -1 & 0 & -1 & -1 \end{bmatrix}$$

If the sought-after branch voltage vector is premultiplied by the mesh-branch incidence matrix, the result is:

$$\begin{bmatrix} 1 & 1 & 1 & 0 & 0 & 0 & 0 & 0 \\ 0 & 0 & -1 & 1 & 1 & 1 & 0 & 0 \\ 0 & -1 & 0 & -1 & -1 & 0 & -1 & -1 \end{bmatrix} \begin{bmatrix} v_1 \\ v_2 \\ v_3 \\ v_4 \\ v_5 \\ v_6 \\ v_7 \\ v_8 \end{bmatrix}$$

$$= \begin{bmatrix} (v_1 + v_2 + v_3) \\ (-v_3 + v_4 + v_5 + v_6) \\ (-v_2 - v_4 - v_5 - v_7 - v_8) \end{bmatrix}$$

If this is compared to an application of the Kirchhoff voltage law for all three meshes, one observes that the algebraic sum of the voltage drops around each closed mesh (voltage rises are negative voltage drops) is equal to zero. Thus

$$\begin{bmatrix} (v_1 + v_2 + v_3) \\ (-v_3 + v_4 + v_5 + v_6) \\ (-v_2 - v_4 - v_5 - v_7 - v_8) \end{bmatrix} = \begin{bmatrix} 0 \\ 0 \\ 0 \end{bmatrix}$$

is a statement of the Kirchhoff voltage law, and for any network

$$MV = 0 \qquad\qquad (9.7)$$

The relationship between mesh and branch currents comes from noting that

$$j_1 = i_1$$

$$j_2 = i_1 - i_3$$

$$j_3 = i_1 - i_2$$

$$j_4 = i_2 - i_3$$

$$j_5 = i_2 - i_3$$

$$j_6 = i_2$$

$$j_7 = -i_3$$

and

$$j_8 = -i_3$$

One may construct a matrix C that transforms the mesh current vector I into the branch current vector J:

$$
\begin{bmatrix}
1 & 0 & 0 \\
1 & 0 & -1 \\
1 & -1 & 0 \\
0 & 1 & -1 \\
0 & 1 & -1 \\
0 & 1 & 0 \\
0 & 0 & -1 \\
0 & 0 & -1
\end{bmatrix}
\begin{bmatrix}
i_1 \\
i_2 \\
i_3
\end{bmatrix}
=
\begin{bmatrix}
j_1 \\
j_2 \\
j_3 \\
j_4 \\
j_5 \\
j_6 \\
j_7 \\
j_8
\end{bmatrix}
$$

and it is seen that $C = M^{\mathrm{T}}$. Thus the mapping from I to J is through the transpose of the mesh-branch incidence matrix:

$$J = M^{\mathrm{T}}I \qquad (9.8)$$

Now take Eq. (9.1):

$$V = RJ + V_s - RJ_s$$

and premultiply every term by M:

$$MV = MRJ + MV_s - MRJ_s$$

Then use Eq. (9.7), $MV = 0$, and Eq. (9.8), $J = M^{\mathrm{T}}I$, to show first that

$$0 = MRJ + MV_s - MRJ_s$$

and then

$$MRM^TI = MRJ_s - MV_s$$

This is the statement of the *mesh equations*:

$$Z_mI = E_s \qquad (9.9)$$

where

$$Z_m = MRM^T \qquad (9.10a)$$

is the *mesh impedance matrix* and

$$E_s = MRJ_s - MV_s \qquad (9.10b)$$

is the *source voltage vector*.

9.2.6 Procedure for Mesh Analysis

The procedure for conducting a mesh analysis also follows a logical sequence:

1. Construct an oriented graph that clearly identifies all branches and nodes. Identify the meshes and insert clockwise mesh currents.
2. Write the mesh-branch incidence matrix M, the current and voltage source vectors J_s and V_s, and the resistance matrix R.
3. Obtain $Z_m = MRM^T$ and $E_s = MRJ_s - MV_s$.
4. By any computationally efficient method, obtain I; for example, $I = Z_m^{-1}E_s$.
5. Determine the branch currents from Eq. (9.8), $J = M^TI$.
6. Determine the branch voltages from Eq. (9.1), $V = RJ + V_s - RJ_s$.

Example: A node-to-datum analysis has been conducted using the network of Fig. 9.4. The result of this node-to-datum analysis is to be verified through the use of a mesh analysis. The oriented graph of Fig. 9.4b is repeated here as Fig. 9.6 with the addition of four clockwise mesh

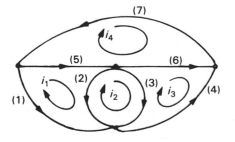

Figure 9.6 An oriented graph of the network of Fig. 9.4a showing four meshes, with mesh currents indicated by the letter *i*.

currents that serve to designate the meshes. The voltage and current source vectors are the same as those used in the node-to-datum analysis:

$$V_s = \begin{bmatrix} 12 \\ 0 \\ 16 \\ 0 \\ 0 \\ 0 \\ 24 \end{bmatrix} \qquad J_s = \begin{bmatrix} -2 \\ 0 \\ 4 \\ 16 \\ 0 \\ 0 \\ 0 \end{bmatrix}$$

and the resistance matrix is:

$$R = \begin{bmatrix} 1 & 0 & 0 & 0 & 0 & 0 & 0 \\ 0 & 8 & 0 & 0 & 0 & 0 & 0 \\ 0 & 0 & 1 & 0 & 0 & 0 & 0 \\ 0 & 0 & 0 & 4 & 0 & 0 & 0 \\ 0 & 0 & 0 & 0 & 2 & 0 & 0 \\ 0 & 0 & 0 & 0 & 0 & 1 & 0 \\ 0 & 0 & 0 & 0 & 0 & 0 & 4 \end{bmatrix}$$

The mesh-branch incidence matrix is seen to be:

$$M = \begin{bmatrix} -1 & 1 & 0 & 0 & 1 & 0 & 0 \\ 0 & -1 & 1 & 0 & 0 & 0 & 0 \\ 0 & 0 & -1 & -1 & 0 & 1 & 0 \\ 0 & 0 & 0 & 0 & -1 & -1 & -1 \end{bmatrix}$$

and the mesh impedance matrix is $Z_m = MRM^T$. First obtain MR, because it is also needed in the formulation of the voltage source vector. The reader can verify that MR is equal to

$$MR = \begin{bmatrix} -1 & 8 & 0 & 0 & 2 & 0 & 0 \\ 0 & -8 & 1 & 0 & 0 & 0 & 0 \\ 0 & 0 & -1 & -4 & 0 & 1 & 0 \\ 0 & 0 & 0 & 0 & -2 & -1 & -4 \end{bmatrix}$$

and then

$$Z_m = \begin{bmatrix} -1 & 8 & 0 & 0 & 2 & 0 & 0 \\ 0 & -8 & 1 & 0 & 0 & 0 & 0 \\ 0 & 0 & -1 & -4 & 0 & 1 & 0 \\ 0 & 0 & 0 & 0 & -2 & -1 & -4 \end{bmatrix} \begin{bmatrix} -1 & 0 & 0 & 0 \\ 1 & -1 & 0 & 0 \\ 0 & 1 & -1 & 0 \\ 0 & 0 & -1 & 0 \\ 1 & 0 & 0 & -1 \\ 0 & 0 & 1 & -1 \\ 0 & 0 & 0 & -1 \end{bmatrix}$$

or

$$Z_m = \begin{bmatrix} 11 & -8 & 0 & -2 \\ -8 & 9 & -1 & 0 \\ 0 & -1 & 6 & -1 \\ -2 & 0 & -1 & 7 \end{bmatrix}$$

The voltage source vector is $E_s = MRJ_s - MV_s$:

$$E_s = \begin{bmatrix} -1 & 8 & 0 & 0 & 2 & 0 & 0 \\ 0 & -8 & 1 & 0 & 0 & 0 & 0 \\ 0 & 0 & -1 & -4 & 0 & 1 & 0 \\ 0 & 0 & 0 & 0 & -2 & -1 & -4 \end{bmatrix} \begin{bmatrix} -2 \\ 0 \\ 4 \\ 16 \\ 0 \\ 0 \\ 0 \end{bmatrix}$$

$$- \begin{bmatrix} -1 & 1 & 0 & 0 & 1 & 0 & 0 \\ 0 & -1 & 1 & 0 & 0 & 0 & 0 \\ 0 & 0 & -1 & -1 & 0 & 1 & 0 \\ 0 & 0 & 0 & 0 & -1 & -1 & -1 \end{bmatrix} \begin{bmatrix} 12 \\ 0 \\ 16 \\ 0 \\ 0 \\ 0 \\ 24 \end{bmatrix}$$

or

$$E_s = \begin{bmatrix} 2 \\ 4 \\ -68 \\ 0 \end{bmatrix} - \begin{bmatrix} -12 \\ 16 \\ -16 \\ -24 \end{bmatrix} = \begin{bmatrix} 14 \\ -12 \\ -52 \\ 24 \end{bmatrix}$$

The mesh equations are $Z_m I = E_s$:

$$\begin{bmatrix} 11 & -8 & 0 & -2 \\ -8 & 9 & -1 & 0 \\ 0 & -1 & 6 & -1 \\ -2 & 0 & -1 & 7 \end{bmatrix} \begin{bmatrix} i_1 \\ i_2 \\ i_3 \\ i_4 \end{bmatrix} = \begin{bmatrix} 14 \\ -12 \\ -52 \\ 24 \end{bmatrix}$$

These are solved using the Crout reduction, with the comparison to the Cholesky reduction and matrix inversion approaches left to the reader if such comparison is desired.

The Crout reduction factors Z_m into lower and upper triangular matrices, $Z_m = LU$. Here, use of Eq. (5.16a) gives:

$$u_{11} = u_{22} = u_{33} = u_{44} = 1.0000$$

with z's indicating the elements of Z_m:

$$l_{11} = z_{11} = \quad 11.0000$$
$$l_{21} = z_{21} = -8.0000$$
$$l_{31} = z_{31} = \quad 0.0000$$
$$l_{41} = z_{41} = -2.0000$$

Then from Eq. (5.16c), with $k = 1$,

$$u_{12} = \frac{1}{11.0000}(-8) = -0.7273$$

$$u_{13} = \frac{1}{11.0000}(0) \quad = 0.0000$$

$$u_{14} = \frac{1}{11.0000}(-2) = -0.1818$$

With $k = 2$ in Eq. (5.16b),

$$l_{22} = 9 - (-8)(-0.7273) = 3.1818$$
$$l_{32} = -1 - (0)(-0.7273) = -1.0000$$
$$l_{42} = 0 - (-2)(-0.7273) = -1.4545$$

and then Eq. (5.16c) provides

$$u_{23} = \frac{1}{3.1818}[-1 - (-8)(0)] = -0.3143$$

$$u_{24} = \frac{1}{3.1818}[0 - (-8)(-0.1818)] = -0.4571$$

Use of Eq. (5.16b) with $k = 3$ yields:

$$l_{33} = 6 - (0)(0) - (-1.0000)(-0.3143) = 5.6857$$

$$l_{43} = -1 - (-2)(0) - (-1.4545)(-0.3143) = -1.4571$$

and then from Eq. (5.16c)

$$u_{34} = \frac{1}{5.6857}[-1 - (0)(-0.1818) - (-1.0000)(-0.4571)] = -0.2563$$

Finally, with $k = 4$ in Eq. (5.16b),

$$l_{44} = 7 - (-2)(-0.1818) - (-1.4545)(-0.4571)$$
$$- (-1.4571)(-0.2563)$$

which gives

$$l_{44} = 5.5980$$

The Crout reduction is LU. The matrix L is

$$\begin{bmatrix} 11.0000 & 0 & 0 & 0 \\ -8.0000 & 3.1818 & 0 & 0 \\ 0 & -1.0000 & 5.6857 & 0 \\ -2.0000 & -1.4545 & -1.4571 & 5.5980 \end{bmatrix}$$

and the matrix U is

$$\begin{bmatrix} 1.0000 & -0.7273 & 0 & -0.1818 \\ 0 & 1.0000 & -0.3413 & -0.4571 \\ 0 & 0 & 1.0000 & -0.2563 \\ 0 & 0 & 0 & 1.0000 \end{bmatrix}$$

and it is easily verified that the foregoing matrix product does indeed equal Z_m.

The back substitution algorithm of Eq. (5.20) can now be used in $LY = B$ with

$$B = \begin{bmatrix} 14 \\ -12 \\ -52 \\ 24 \end{bmatrix}$$

The elements of Y are:

$$y_1 = \frac{1}{11.0000} (14) = 1.2727$$

$$y_2 = \frac{1}{3.1818} [-12 - (-8)(1.2727)] = -0.5714$$

$$y_3 = \frac{1}{5.6867} [-52 - (0)(1.2727) - (-1)(-0.5714)] = -9.2462$$

and

$$y_4 = \frac{1}{5.5980} [24 - (-2)(1.2727) - (-1.4545)(-0.5714)$$

$$- (-1.4571)(-9.2462)]$$

or

$$y_4 = 2.1867$$

Then via Eq. (5.21) the elements of I are obtained. Notice that all elements on the principal diagonal of U are unity so that

$$i_4 = 2.1867$$

$$i_3 = -9.2462 - (-0.2563)(2.1867) = -8.6858$$

$$i_2 = -0.5714 - (-0.3143)(-8.6858)$$

$$- (-0.4571)(2.1867) = -2.3106$$

and

$$i_1 = 1.2727 - (-0.7273)(-2.3106)$$

$$- (0)(-8.6858) - (-0.1818)(2.1867)$$

or

$$i_1 = -0.0036$$

Thus the mesh current vector (in amperes) is:

$$I = \begin{bmatrix} -0.0036 \\ -2.3106 \\ -8.6858 \\ 2.1867 \end{bmatrix}$$

and the branch currents (in amperes) can be determined through the use of Eq. (9.8), $J = M^T I$

$$J = \begin{bmatrix} -1 & 0 & 0 & 0 \\ 1 & -1 & 0 & 0 \\ 0 & 1 & -1 & 0 \\ 0 & 0 & -1 & 0 \\ 1 & 0 & 0 & -1 \\ 0 & 0 & 1 & -1 \\ 0 & 0 & 0 & -1 \end{bmatrix} \begin{bmatrix} -0.0036 \\ -2.3106 \\ -8.6858 \\ 2.1867 \end{bmatrix} = \begin{bmatrix} 0.0036 \\ 2.2980 \\ 6.3842 \\ 8.6858 \\ -2.1903 \\ -10.8725 \\ -2.1867 \end{bmatrix}$$

These branch currents agree, as they should, with the branch currents obtained in the node-to-datum analysis.

The branch voltages (in volts) are found from an application of Eq. (9.1), $V = RJ + V_s - RJ_s$. First,

$$RJ = \begin{bmatrix} 1 & 0 & 0 & 0 & 0 & 0 & 0 \\ 0 & 8 & 0 & 0 & 0 & 0 & 0 \\ 0 & 0 & 1 & 0 & 0 & 0 & 0 \\ 0 & 0 & 0 & 4 & 0 & 0 & 0 \\ 0 & 0 & 0 & 0 & 2 & 0 & 0 \\ 0 & 0 & 0 & 0 & 0 & 1 & 0 \\ 0 & 0 & 0 & 0 & 0 & 0 & 4 \end{bmatrix} \begin{bmatrix} 0.0036 \\ 2.2980 \\ 6.3842 \\ 8.6858 \\ -2.9103 \\ -10.8725 \\ -2.1867 \end{bmatrix} = \begin{bmatrix} 0.0036 \\ 18.3842 \\ 6.3842 \\ 37.7433 \\ -4.3806 \\ -10.8725 \\ -8.7469 \end{bmatrix}$$

so that with a simple matrix multiplication for RJ_s

$$
V = \begin{bmatrix} 0.0036 \\ 18.3842 \\ 6.3842 \\ 37.7433 \\ -4.3806 \\ -10.8725 \\ -8.7409 \end{bmatrix} + \begin{bmatrix} 12 \\ 0 \\ 16 \\ 0 \\ 0 \\ 0 \\ 24 \end{bmatrix} - \begin{bmatrix} -2 \\ 0 \\ 4 \\ 64 \\ 0 \\ 0 \\ 0 \end{bmatrix} = \begin{bmatrix} 14.0036 \\ 18.3842 \\ 18.3842 \\ -29.2567 \\ -4.3806 \\ -10.8725 \\ 15.2531 \end{bmatrix}
$$

This, too, is identical to the result obtained in the node-to-datum analysis.

9.3 STATE SPACE OR STATE VARIABLE ANALYSIS

9.3.1 Preliminaries

Modern engineering systems can be quite complex. They may be subjected to a variety of simultaneous inputs, and it is possible that more than one output can be of interest. Moreover, the system input and output, whether single or multiple, may be interrelated in a complicated fashion that poses definite problems with regard to system analysis. State space analysis has the ability to focus on the variables of interest, reduce the complexity of the mathematical analysis, and permit the use of a computer for the bulk of the computations necessary for the successful completion of the analysis.

9.3.2 Definitions

The state of a system pertains to the minimum set of quantities or variables that, because they contain enough data or information regarding the history of the system, enable the future performance of the system to be predicted. The variables themselves are called the *state variables*.

The state variables are designated as $x_1(t), x_2(t), \ldots, x_3(t)$ even though, if considered separately, they appear to bear little resemblance to one another. But if at any time t_0, the variables possess values $x_1(t_0), x_2(t_0), \ldots, x_n(t_0)$, the initial state is considered as specified. For $t > t_0$, if an input is specified, the state variables completely define the behavior of the system.

If the *n*-state variables are assembled into an $n \times 1$ column vector, the vector is called the *state vector*, and each element of the state vector is a state variable. The state vector represents the system state for all $t > t_0$ if the system input or perturbation is specified and if the initial state of the system is known. The n-dimensional space with coordinate axes x_1, x_2, . . ., x_n is called the *state space*. Any state can be represented as a point in this state space. Analyses conducted using the concept of state is called state space analysis or state variable analysis.

9.3.3 The State Space Formulation

In Section 8.10 it was shown that an nth-order ordinary linear differential equation with dependent variable y and independent variable t could be represented as a set of n simultaneous linear first-order differential equations. The set of first-order differential equations is the state space representation.

Example: Consider the second-order differential equation discussed in Section 8.11:

$$\frac{d^2y}{dt^2} + 5\frac{dy}{dt} + 6y = 16$$

with $y(0) = 0$ and $dy/dt|_{t=0} = 4$

Here, two state variables may be proposed, $x_1 = y$ and $x_2 = \dot{y}$ where the dot rather than a prime indicates the derivative with respect to time. The derivative of these may be taken so that $\dot{x}_1 = x_2 = \dot{y}$ and $\dot{x}_2 = \ddot{y}$, and this allows the second-order equation to be written as

$$\dot{x}_1 = x_2$$

$$\dot{x}_2 = -6x_1 - 5x_2 + 16$$

In matrix form these become

$$\begin{bmatrix} \dot{x}_1 \\ \dot{x}_2 \end{bmatrix} = \begin{bmatrix} 0 & 1 \\ -6 & -5 \end{bmatrix} \begin{bmatrix} x_1 \\ x_2 \end{bmatrix} + \begin{bmatrix} 0 \\ 16 \end{bmatrix} u(t)$$

where $u(t)$ is called the *unit step function*. The form

$$X = AX + BR \tag{9.11}$$

is called the state space or state variable formulation or form. Here, in this specific case, the two state variables are $x_1(t)$ and $x_2(t)$, but in general, n-state variables form the state vector. The $n \times n$ matrix A is called the *coefficient matrix*.

The matrix R is the *input vector*, and if there are p different inputs, R is $p \times 1$. In the event that only one input is provided, or if all inputs are of the same form, $p = 1$ and R reduces to the scalar $r(t)$, and the matrix B is $n \times p$.

The solution of the system of Eq. (9.11) is subject to a set of initial conditions that describe the state of the system. These conditions are contained in an $n \times 1$ initial condition vector X_0:

$$X_0 = \begin{bmatrix} x_1(t_0) \\ x_2(t_0) \\ x_3(t_0) \\ \cdots \\ x_n(t_0) \end{bmatrix}$$

or if $t_0 = 0$, which is most often the case,

$$X_0 = \begin{bmatrix} x_1(0) \\ x_2(0) \\ x_3(0) \\ \cdots \\ x_n(0) \end{bmatrix}$$

The procedure for the solution of Eq. (9.11) has been given in Section 8.11.

9.3.4 Two Examples of State Space Formulations

This section considers two examples of the formulation of the state space or state variable problem.

Example: The torsional mechanical system shown in Fig. 9.7 possesses a flywheel with moment of inertia J and a shaft with stiffness k. The bearing provides viscous drag through a damping coefficient c that is proportional to the shaft angular velocity. At $t = 0$, the shaft displacement $\theta = 0$, the shaft angular velocity $\omega = 0$ and the system is suddenly subjected to an external torque T.

When the external torque is applied as shown in Fig. 9.7, the system tends to rotate in a clockwise direction. This rotation is resisted by a restraining torque due to the bearing and the shaft. Newton's law equates all the external torques to the product of the system moment of inertia and the angular acceleration. Thus the differential equation of motion is:

$$T - c\omega - k\theta = J\alpha$$

or

$$J\alpha + c\omega + k\theta = T$$

Here the state variables are the angular displacement θ and the angular

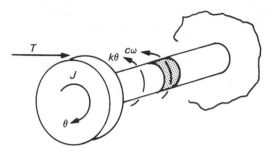

Figure 9.7 Mechanical rotational system.

velocity $\omega = d\theta/dt$. Thus $x_1 = \theta$ and $x_2 = \omega$ and

$$\dot{x}_1 = \frac{d\theta}{dt} = \omega$$

and with

$$\alpha = -\frac{k}{J}\theta - \frac{c}{J}\omega + \frac{T}{J}$$

or

$$\dot{x}_2 = \frac{d\omega}{dt} = \alpha = -\frac{k}{J}x_1 - \frac{c}{J}x_2 + \frac{T}{J}$$

it is easy to see that the state variable formulation is:

$$\begin{bmatrix} \dot{x}_1 \\ \dot{x}_2 \end{bmatrix} = \begin{bmatrix} 0 & 1 \\ -k/J & -c/J \end{bmatrix} \begin{bmatrix} x_1 \\ x_2 \end{bmatrix} + \begin{bmatrix} 0 \\ T/J \end{bmatrix} u(t)$$

where $u(t)$ is the unit step function. The initial condition vector in this case is:

$$X_0 = \begin{bmatrix} 0 \\ 0 \end{bmatrix}$$

Example: The electrical system shown in Fig. 9.8a contains two voltage sources, two capacitors, and two inductors. The switches S_1 and S_2 close at $t = 0$ when current flows in either inductor and capacitor C_1 is charged to V_0 volts.

A knowledge of the currents through each of the inductors and the voltages across both of the capacitors is all that is needed to describe completely the state of this network. Thus the state variables may be designated as $x_1 = i_{L1}$, $x_2 = i_{L2}$, $x_3 = v_{c1}$, and $x_4 = v_{c2}$. This makes the initial condition vector:

$$X_0 = \begin{bmatrix} 0 \\ 0 \\ V_0 \\ 0 \end{bmatrix}$$

Two of the state equations are obtained through an application of Kirchhoff's voltage law around the two meshes in Fig. 9.8*b*. The other two state equations derive from a consideration of Kirchhoff's current law at nodes 1 and 2 designated by numerals within circles in Fig. 9.8*b*. For the meshes:

$$R_1(i_1 - i_2) + L_1 \frac{di_1}{dt} + v_1 = E_1$$

and

$$L_2 \frac{di_2}{dt} + R_2 i_2 + v_2 - v_1 + R_1(i_2 - i_1) = E_2$$

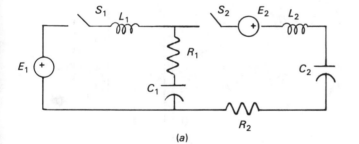

(a)

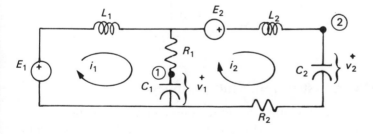

(b)

Figure 9.8 (*a*) An electrical network and (*b*) the network showing the state variables.

Then for the nodes:

$$C_1 \frac{dv_1}{dt} = i_1 - i_2$$

and

$$C_2 \frac{dv_2}{dt} = i_2$$

The foregoing equations may be arranged to express the derivatives of the state variables in terms of the other quantities:

$$\frac{di_1}{dt} = -\frac{R_1}{L_1} i_1 + \frac{R_1}{L_1} i_2 - \frac{1}{L_1} v_1 + \frac{1}{L_1} E_1$$

$$\frac{di_2}{dt} = \frac{R_1}{L_2} i_1 - \frac{R_1 + R_2}{L_2} i_2 + \frac{1}{L_2} v_1 - \frac{1}{L_2} v_2 + \frac{1}{L_2} E_2$$

$$\frac{dv_1}{dt} = \frac{1}{C_1} i_1 - \frac{1}{C_2} i_2$$

$$\frac{dv_2}{dt} = \frac{1}{C_2} i_2$$

and the state variable equations are:

$$
\begin{bmatrix} \dfrac{di_1}{dt} \\[2mm] \dfrac{di_2}{dt} \\[2mm] \dfrac{dv_1}{dt} \\[2mm] \dfrac{dv_2}{dt} \end{bmatrix}
=
\begin{bmatrix}
-\dfrac{R_1}{L_1} & \dfrac{R_1}{L_1} & -\dfrac{1}{L_1} & 0 \\[2mm]
\dfrac{R_1}{L_2} & -\dfrac{R_1 + R_2}{L_2} & +\dfrac{1}{L_2} & -\dfrac{1}{L_2} \\[2mm]
\dfrac{1}{C_1} & -\dfrac{1}{C_2} & 0 & 0 \\[2mm]
0 & \dfrac{1}{C_2} & 0 & 0
\end{bmatrix}
\begin{bmatrix} i_1 \\[2mm] i_2 \\[2mm] v_1 \\[2mm] v_2 \end{bmatrix}
+
\begin{bmatrix} \dfrac{E_1}{L_1} \\[2mm] \dfrac{E_2}{L_2} \\[2mm] 0 \\[2mm] 0 \end{bmatrix}
u(t)
$$

9.3.5 A Feedback System with Multiple Inputs

A feedback system containing two integrating controllers is displayed in block diagram form in Fig. 9.9. The system can be perturbed by step inputs at two locations, and these perturbations are indicated in the form of unit step inputs with strengths of 4 and 2. The state variables are x_1 and x_2, and the output of the system is $x_1 = y$. The system is at rest at the instant

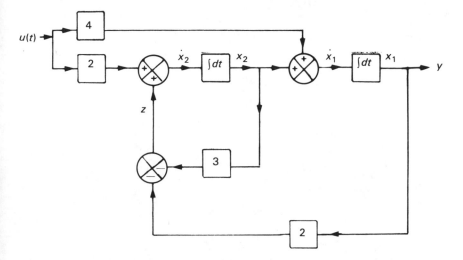

Figure 9.9 Block diagram of feedback system.

that the perturbations are impressed so that the initial condition vector is:

$$X_0 = \begin{bmatrix} 0 \\ 0 \end{bmatrix}$$

A knowledge of the algorithms for block diagram reduction is not necessary in this case, because the state equations can be obtained from an application of continuity at each summation point. It is easy to see that the state equations are:

$$\dot{x}_2 = z + 2u(t)$$

$$\dot{x}_1 = x_2 + 4u(t)$$

with

$$z = -2x_1 - 3x_2$$

so that

$$\dot{x}_1 = x_2 + 4u(t)$$

$$\dot{x}_2 = -2x_1 - 3x_2 + 2u(t)$$

In the state variable form of Eq. (9.11), these become:

$$\begin{bmatrix} \dot{x}_1 \\ \dot{x}_2 \end{bmatrix} = \begin{bmatrix} 0 & 1 \\ -2 & -3 \end{bmatrix} \begin{bmatrix} x_1 \\ x_2 \end{bmatrix} = \begin{bmatrix} 4 \\ 2 \end{bmatrix} u(t)$$

with

$$X_0 = \begin{bmatrix} x_1(0) \\ x_2(0) \end{bmatrix} = \begin{bmatrix} 0 \\ 0 \end{bmatrix}$$

Observe that in a comparison to Eq. (9.11):

$$\dot{X} = AX + BR$$

the vector R is a scalar:

$$R = u(t)$$

and B is a column vector:

$$B = \begin{bmatrix} 4 \\ 2 \end{bmatrix}$$

The solution procedure of Section 8.11 can be followed:

$$[A - \lambda I] = \begin{bmatrix} -\lambda & 1 \\ -2 & -\lambda - 3 \end{bmatrix}$$

so that

$$p(\lambda) = \det [A - \lambda I] = \lambda^2 + 3\lambda + 2 = 0$$

This makes the eigenvalues:

$$\lambda_1 = -2$$

$$\lambda_2 = -1$$

The eigenvector associated with $\lambda_1 = -2$ derives from a solution of the following system:

$$\begin{bmatrix} 2 & 1 \\ -2 & -1 \end{bmatrix} \begin{bmatrix} x_1 \\ x_2 \end{bmatrix} = \begin{bmatrix} 0 \\ 0 \end{bmatrix}$$

which indicates that the eigenvector is:

$$X_1 = \begin{bmatrix} 1 \\ -2 \end{bmatrix}$$

The second eigenvector corresponding to the eigenvalue $\lambda_2 = -1$ is obtained from the following system:

$$\begin{bmatrix} 1 & 1 \\ -2 & -2 \end{bmatrix} \begin{bmatrix} x_1 \\ x_2 \end{bmatrix} = \begin{bmatrix} 0 \\ 0 \end{bmatrix}$$

so that the eigenvector is:

$$X_2 = \begin{bmatrix} 1 \\ -1 \end{bmatrix}$$

The eigenvectors are linearly independent, and the modal matrix is:

$$M = \begin{bmatrix} 1 & 1 \\ -2 & -1 \end{bmatrix}$$

and this has an inverse:

$$M^{-1} = \begin{bmatrix} -1 & -1 \\ 2 & 1 \end{bmatrix}$$

An application of Eq. (8.18) provides e^{At}

$$e^{At} = Mf(\lambda)M^{-1}$$

Thus

$$e^{At} = \begin{bmatrix} 1 & 1 \\ -2 & -1 \end{bmatrix} \begin{bmatrix} e^{-2t} & 0 \\ 0 & e^{-t} \end{bmatrix} \begin{bmatrix} -1 & -1 \\ 2 & 1 \end{bmatrix}$$

$$= \begin{bmatrix} e^{-2t} & e^{-t} \\ -2e^{-2t} & -e^{-t} \end{bmatrix} \begin{bmatrix} -1 & -1 \\ 2 & 1 \end{bmatrix}$$

or

$$e^{At} = \begin{bmatrix} (-e^{-2t} + 2e^{-t}) & (-e^{-2t} + e^{-t}) \\ (2e^{-2t} - 2e^{-t}) & (2e^{-2t} - e^{-t}) \end{bmatrix}$$

The solution is obtained from en employment of Eq. (8.22) modified in accordance with the nomenclature of this section:

$$X = e^{At}X_0 + \int_0^t e^{A(t-z)}Bu(z) \, dz$$

Here X_0 is null ($X_0 = 0$) and $u(z) = 1$. Thus

$$X = \int_0^t e^{A(t-z)}B \, dz$$

and the product $e^{A(t-z)}B$ can be evaluated:

$$e^{A(t-z)}B = \begin{bmatrix} (-e^{-2(t-z)} + 2e^{-(t-z)}) & (-e^{-2(t-z)} + e^{-(t-z)}) \\ (2e^{-2(t-z)} - 2e^{-(t-z)}) & (2e^{-2(t-z)} - e^{-(t-z)}) \end{bmatrix} \begin{bmatrix} 4 \\ 2 \end{bmatrix}$$

or

$$e^{A(t-z)}B = \begin{bmatrix} (-6e^{-2(t-z)} + 10e^{-(t-z)}) \\ (12e^{-2(t-z)} - 10e^{-(t-z)}) \end{bmatrix}$$

Then the integration may be performed

$$\int_0^t e^{A(t-z)}B\,dz = \int_0^t \begin{bmatrix} (-6e^{-2(t-z)} + 10e^{-(t-z)}) \\ (12e^{-2(t-z)} - 10e^{-(t-z)}) \end{bmatrix} dz$$

$$= \begin{bmatrix} (-3e^{-2(t-z)} + 10e^{-(t-z)}) \\ (6e^{-2(t-z)} - 10e^{-(t-z)}) \end{bmatrix}_0^t$$

or

$$X = \begin{bmatrix} 3e^{-2t} - 10e^{-t} + 7 \\ -6e^{-2t} + 10e^{-t} - 4 \end{bmatrix}$$

The output of the system is $y = x_1$:

$$y = 3e^{-2t} - 10e^{-t} + 7$$

9.4 UNDAMPED FREE VIBRATIONS OF A MECHANICAL SYSTEM

9.4.1 Problem Formulation

A system of n masses connected by n springs, but with no damping, constitutes a multi-degree-of-freedom (an n-degree-of-freedom) conservative system. In what follows, three masses and springs are used to describe a three-degree-of-freedom system. The analysis techniques developed for the three-degree-of-freedom system can be extended, if desired to the n-degree-of-freedom system.

The system of three masses connected by springs shown in Fig. 9.10a is constrained to move with translational motion, and the displacement of each mass with respect to an at-rest origin is measured along the length coordinates x_1, x_2, and x_3. Here there is no damping, so that the system is conservative, and because of this, it is evident that each mass must vibrate about its at-rest or equilibrium position (the origin of each length coordinate) with constant amplitude.

Newton's law may be applied to each of the masses to establish the equations of motion. It is easy to see from each of the free body diagrams in Fig. 9.10b that, with dots used to indicate derivatives with respect to

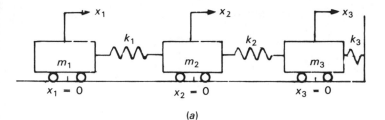

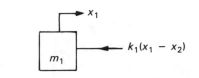

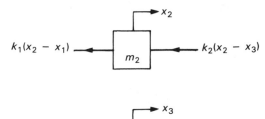

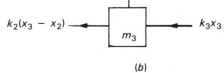

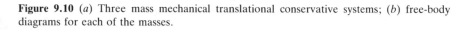

(b)

Figure 9.10 (a) Three mass mechanical translational conservative systems; (b) free-body diagrams for each of the masses.

time,

$$m_1\ddot{x}_1 + k_1(x_1 - x_2) = 0$$
$$m_2\ddot{x}_2 + k_1(x_2 - x_1) + k_2(x_2 - x_3) = 0$$
$$m_3\ddot{x}_3 + k_2(x_3 - x_2) + k_3x_3 = 0$$

A simple rearrangement gives the following:

$$m_1\ddot{x}_1 + k_1x_1 - k_1x_2 = 0$$
$$m_2\ddot{x}_2 - k_1x_1 + (k_1 + k_2)x_2 - k_2x_3 = 0$$
$$m_3\ddot{x}_3 - k_2x_2 + (k_2 + k_3)x_3 = 0$$

In matrix form, these can be represented by:

$$M\ddot{X} + KX = 0 \tag{9.12}$$

where $\ddot{X}$ and X are 3×1 column vectors:

$$\ddot{X} = \begin{bmatrix} \ddot{x}_1 \\ \ddot{x}_2 \\ \ddot{x}_3 \end{bmatrix} ; \qquad X = \begin{bmatrix} x_1 \\ x_2 \\ x_3 \end{bmatrix}$$

M is a 3×3 inertia matrix:

$$M = \begin{bmatrix} m_1 & 0 & 0 \\ 0 & m_2 & 0 \\ 0 & 0 & m_3 \end{bmatrix}$$

and K is a 3×3 stiffness matrix:

$$K = \begin{bmatrix} k_1 & -k_1 & 0 \\ -k_1 & (k_1 + k_2) & -k_2 \\ 0 & -k_2 & (k_2 + k_3) \end{bmatrix}$$

9.4.2 The Eigenvalue Problem and the Vibration Modes

Premultiplication of Eq. (9.12) by the inverse of K (K is often referred to as the *flexibility matrix*) yields $K^{-1}M\ddot{X} + X = 0$ or

$$D\ddot{X} + X = 0 \qquad (9.13a)$$

where $D = K^{-1}M$ is called the *dynamical matrix*.

On the other hand, because the inertia matrix is diagonal, a premultiplication of Eq. (9.12) by M^{-1}, which is a diagonal matrix whose diagonal elements consist of the reciprocals of the diagonal elements of M, gives $\ddot{X} + M^{-1}KX = 0$ or

$$\ddot{X} + WX = 0 \qquad (9.13b)$$

Here W is called the *inverse dynamical matrix*. Notice that

$$W = M^{-1}K = D^{-1} = (K^{-1}M)^{-1} = M^{-1}K$$

which shows that the nomenclature describing D and W is meaningful.

In the conservative system of Fig. 9.10a, each mass must vibrate about its equilibrium position with constant amplitude:

$$X = A \cos \omega t + B \sin \omega t$$

where A and B are 3×1 amplitude vectors. It is simpler, however, to represent this motion in the form of an amplitude and phase angle:

$$X = C \cos (\omega t + \phi)$$

where C is a 3×1 column vector. Thus

$$\ddot{X} = -\omega^2 C \cos(\omega t + \phi)$$

and with these representations put into Eq. (9.13b), the result is:

$$-\omega^2 C + WC = 0$$

or

$$WC = \lambda C \qquad (9.14)$$

where

$$\lambda = \omega^2 \qquad (9.15)$$

This is the eigenvalue problem, and the matrix W has a characteristic equation:

$$p(\lambda) = \det[W - \lambda I] = 0$$

Thus the natural frequencies can be easily obtained by extracting the square root of each eigenvalue (discarding the negative frequency):

$$\omega_i = \sqrt{\lambda_i} \qquad i = 1, 2, 3 \qquad (9.16)$$

Each eigenvalue possesses an associated eigenvector that satisfies

$$WC_i = \lambda C_i \qquad i = 1, 2, 3$$

where the eigenvectors C_i are 3×1. The eigenvectors can be assembled into a modal matrix P (the use of M for the modal matrix may lead to confusion with the inertia matrix):

$$P = [C_1 \quad C_2 \quad C_3]$$

The eigenvectors represent the natural vibration modes. They also possess an orthogonality with respect to both the inertia matrix and the stiffness matrix K. This property depends on both M and K being symmetric.

Consider C_1 and C_2. Both of these must satisfy Eq. (9.14) with $\lambda = \omega^2$ as indicated by Eq. (9.15):

$$WC_1 = M^{-1}KC_1 = \omega_1^2 C_1$$

$$WC_2 = M^{-1}KC_2 = \omega_2^2 C_2$$

If these are premultiplied by M, the result is:

$$KC_1 = \omega_1^2 MC_1$$

$$KC_2 = \omega_2^2 MC_2$$

Then a premultiplication of the first by C_2^T and the second by C_1^T gives

$$\omega_1^2 C_2^T M C_1 = C_2^T K C_1$$

$$\omega_2^2 C_1^T M C_2 = C_1^T K C_2$$

With both M and K symmetric, the second of these may be transposed and subtracted from the first to give

$$(\omega_1^2 - \omega_2^2) C_1^T M C_2 = 0$$

and because the natural frequencies are not equal, it is seen that the orthogonality condition for the eigenvectors is:

$$C_j^T M C_k = 0 \tag{9.17a}$$

and it is also apparent that another orthogonality condition is:

$$C_k^T K C_j = 0 \tag{9.17b}$$

where j and k are any two eigenvectors that form two of the columns of the modal matrix.

Example: In the two-mass system of Fig. 9.11, $m_1 = 4$ kg and $m_2 = 1$ kg. The values of the spring constants are $k_1 = k_2 = 40$ N/m. Prior to $t = 0$, the mass m_1 is held at a point where $x_1 = 0.10$ m and mass m_2 is constrained to remain at its equilibrium position. At $t = 0$, all constraints are removed, and both masses then begin to oscillate. The foregoing indicates that the four initial conditions can be represented by

$$X_0 = X(0) = \begin{bmatrix} 0.1000 \\ 0 \end{bmatrix} \qquad \dot{X}_0 = \dot{X}(0) = \begin{bmatrix} 0 \\ 0 \end{bmatrix}$$

The inertia and stiffness matrices for this two-mass system are:

$$M = \begin{bmatrix} 4 & 0 \\ 0 & 1 \end{bmatrix}$$

and

$$K = \begin{bmatrix} 40 & -40 \\ -40 & 80 \end{bmatrix}$$

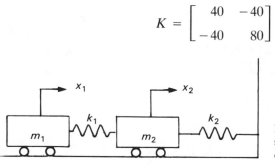

Figure 9.11 Two-mass, two-spring mechanical translational conservative system.

so that the inverse dynamical matrix is:

$$W = \begin{bmatrix} 1/4 & 0 \\ 0 & 1 \end{bmatrix} \begin{bmatrix} 40 & -40 \\ -40 & 80 \end{bmatrix} = \begin{bmatrix} 10 & -10 \\ -40 & 80 \end{bmatrix}$$

The characteristic equation of the matrix W is:

$$p(\lambda) = \lambda^2 - 90\lambda + 400 = 0$$

so that the eigenvalues are:

$$\lambda_1 = 85.3113$$

$$\lambda_2 = 4.6887$$

and the natural frequencies are obtained from $\omega_i = \sqrt{\lambda_i}$:

$$\omega_1 = 9.2364$$

$$\omega_2 = 2.1653$$

which are both in radians/second.

The reader may wish to verify that the two eigenvectors are:

$$C_1 = \begin{bmatrix} 1.0000 \\ -7.5311 \end{bmatrix}$$

corresponding to $\lambda_1 = 85.3113$ ($\omega_1 = 9.2364$ rad/s) and

$$C_2 = \begin{bmatrix} 1.0000 \\ 0.5311 \end{bmatrix}$$

corresponding to $\lambda_2 = 4.6887$ ($\omega_2 = 2.1653$ rad/s).

The eigenvectors are orthogonal to both M and K. From Eq. (9.17a)

$$[1.0000 \quad -7.5311] \begin{bmatrix} 4 & 0 \\ 0 & 1 \end{bmatrix} \begin{bmatrix} 1.0000 \\ 0.5311 \end{bmatrix} = [1.0000 \quad -7.5311] \begin{bmatrix} 4.0000 \\ 0.5311 \end{bmatrix} = 0$$

The two free vibration modes are shown in Fig. 9.12, and it should be noted that the first mode possesses a sign change indicating that at some point between m_1 and m_2 there is no displacement. This point is called a *node*.

The time domain solutions for the motion of masses m_1 and m_2 are:

$$X = \alpha \begin{bmatrix} 1.0000 \\ -7.5311 \end{bmatrix} \cos (9.2364t + \phi) + \beta \begin{bmatrix} 1.0000 \\ 0.5311 \end{bmatrix} \cos (2.1653t + \phi)$$

where the arbitrary constants α and β are obtained from an application of the initial conditions.

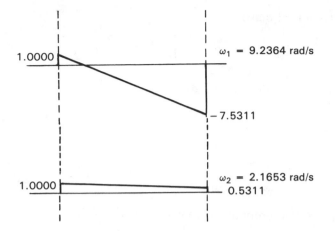

Figure 9.12 Free vibration modes for the two-mass, two-spring system of Fig. 9.11.

9.5 CONCLUSION

This chapter has provided three examples of the use of matrices as a tool to assist in obtaining solutions to problems in engineering analysis. Space does not permit a further demonstration of the power of matrix algebra and calculus in such diverse arenas as structural analysis, the theory of elasticity, electrical two-port and transmission-line analysis, and classical mechanics.

THE ROOTS OF POLYNOMIAL EQUATIONS

A.1 PRELIMINARIES

Define a polynomial equation in x:

$$f(x) = a_n x^n + a_{n-1} x^{n-1} + a_{n-2} x^{n-2} + \cdots + a_1 x_1 + a_0 = 0 \quad (A.1)$$

where n is a positive integer and the a's are real constants. The fundamental theorem of algebra states that every polynomial equation of degree 1 or higher has a root r in the set of complex numbers such that $f(r) = 0$ no matter how high the degree of the equation (and even if the a's are complex constants).

   If the polynomial equation is divided by a divisor of degree 1 or higher that is in itself a function of x, a remainder $R(x)$ is obtained. This is a statement of the remainder theorem and says that if the divisor is $D(x)$, then a quotient polynomial and a remainder come forth:

$$\frac{f(x)}{D(x)} = Q(x) + \frac{R(x)}{D(x)}$$

or

$$f(x) = D(x)Q(x) + R(x)$$

and $R(x)$ is of lesser degree than $D(x)$. In particular, if $D(x)$ is of first degree, $R(x)$ is a constant. If $b = r$ where r is a root of Eq. (A.1), then if $D(x) = x - r$, $R(x) = 0$ and $Q(x)$ is of lesser degree than $f(x)$. In this case, $Q(x)$ is known as the *depressed equation*.

   Every polynomial equation in the form of Eq. (A.1) of degree 1 or higher can, if $a_0 \neq 0$, be expressed as the product of n linear factors of the form $(x - r_k)$ with $k = 1, 2, 3, \ldots, n - 1, n$:

$$f(x) = a_n(x - r_1)(x - r_2)(x - r_3) \cdots (x - r_{n-1})(x - r_n) \quad \text{(A.2)}$$

where the n values of r_k are the roots of $f(x) = 0$. If $(x - r_1)$ is a factor of $f(x)$, then $R(r_1) = 0$ and

$$f(x) = (x - r_1)Q_1(x)$$

where

$$Q_1(x) = a_n x^{n-1} + a_{n-1}x^{n-2} + a_{n-2}x^{n-3} + \cdots$$

However, $x = r_2$ is also a root of $f(x)$ and must therefore be a root of $Q_1(x)$. Thus the use of the factor $(x - r_2)$ yields

$$Q_1(x) = (x - r_2)Q_2(x)$$

where $R(r_2) = 0$ and where $Q_2(x)$ is another depressed equation. It is easy to see that this process may be continued until $Q_n(x)$, which will be of first degree, is reached.

The foregoing suggests that the roots of a polynomial equation in the form of Eq. (A.1) can be found by assuming their value, checking to see if $f(r) = 0$ (which indicates that there is no remainder), and then depressing the equation to $Q(x) = 0$ by dividing $f(x)$ by the factor $(x - r)$. This is a valid technique, but much labor can be saved if one knows where to begin.

A.2 WHERE TO BEGIN AND DESCARTES'S RULE OF SIGNS

Descartes's rule of signs is presented here without formal proof. The rule has several ramifications and pertains to Eq. (A.1):

1. The maximum number of real roots of $f(x) = 0$ cannot exceed the number of variations in sign of $f(x)$, and if the number of positive roots is less than the maximum, the number must be less by an even number.
2. The maximum number of negative real roots of $f(x) = 0$ cannot exceed the number of changes of sign of $f(-x) = 0$, and if the number of negative roots is less than the maximum, the number must be less by an even number.
3. Complex roots must occur in conjugate pairs.
4. If $f(x) = 0$ is of odd degree, at least one real root must exist.

If it is desired to find the roots of a polynomial equation by substituting assumed root values into $f(x) = 0$, Descartes's rule of signs can save a considerable amount of labor. Unfortunately, the statement that complex conjugate roots must occur in pairs gives no information as to whether the real parts of such roots are positive or negative. Furthermore, it should be kept in mind that, in general, $f(-x) \neq -f(x)$.

Example:
Find these roots of

$$f(x) = x^5 + 11x^4 + 49x^3 + 121x^2 + 178x + 120 = 0$$

Solution: First write $f(-x)$:

$$f(-x) = -x^5 + 11x^4 - 49x^3 + 121x^2 - 178x + 120 = 0$$

and then consider $f(x)$ and $f(-x)$. There are no sign changes in $f(x)$, so there can be no positive real roots. However, there are five sign changes in $f(-x)$, so that there can be 5, 3, or 1 negative real roots. This means that there can be 2 or 4 complex roots that must occur in conjugate pairs, and it is indeed a fact that because the degree of $f(x)$ is odd, there must be at least one real root.

Begin by making some guesses as to root values:

$$f(-1) = -1 + 11 - 49 + 121 - 178 + 120 = 24$$

and this shows that -1 is not a root. Then consider

$$f(-2) = -32 + 176 - 392 + 484 - 356 + 120 = 0$$

so that $x = -2$ is a root. One might try $x = -3$:

$$f(-3) = -243 + 891 - 1323 + 1089 - 534 + 120 = 0$$

and find that $x = -3$ is also a root.

Two roots have been found, $x = -2$ and $x = -3$. These form factors $(x + 2)$ and $(x + 3)$, and the product of these factors is $x^2 + 5x + 6$ which is of degree two. If $D(x) = x^2 + 5x + 6$, the original fifth-degree equation can be depressed to a third-degree equation $Q(x)$ with no remainder:

$$
\begin{array}{r}
x^3 + 6x^2 + 13x + 20 \\
x^2 + 5x + 6 \overline{\smash{\big)}\ x^5 + 11x^4 + 49x^3 + 121x^2 + 178x + 120} \\
\underline{x^5 + 5x^4 + 6x^3} \\
6x^4 + 43x^3 + 121x^2 \\
\underline{6x^4 + 30x^3 + 36x^2} \\
13x^3 + 85x^2 + 178x \\
\underline{13x^3 + 65x^2 + 78x} \\
20x^2 + 100x + 120 \\
\underline{20x^2 + 100x + 120} \\
0
\end{array}
$$

The same procedure can be applied to the depressed equation:

$$Q(x) = x^3 + 6x^2 + 13x + 20 = 0$$

Assume $x = -1, -2, -3$, and -4 as necessary:

$$f(-1) = -1 + 6 - 13 + 20 = 12$$
$$f(-2) = -8 + 24 - 26 + 20 = 10$$
$$f(-3) = -27 + 54 - 39 + 20 = 8$$
$$f(-4) = -64 + 96 - 52 + 20 = 0$$

and find that $x = -4$ is a root, $x + 4$ is a factor, and a further depression using $D(x) = x + 4$ yields:

$$
\begin{array}{r}
x^2 + 2x + 5 \\
x + 4 \enclose{longdiv}{x^3 + 6x^2 + 13x + 20} \\
\underline{x^3 + 4x^2} \\
2x^2 + 13x \\
\underline{2x^2 + 8x} \\
5x + 20 \\
\underline{5x + 20} \\
0
\end{array}
$$

The depressed equation

$$Q(x) = x^2 + 2x + 5$$

is a quadratic that is easily solved and the remainder is equal to zero. Thus three of the roots of $f(x)$ are $x_1 = -2$, $x_2 = -3$ and $x_3 = -4$ and the last two come from a solution of $x^2 + 2x + 5 = 0$ so that $x_4, x_5 = -1 \pm j2$ where $j = \sqrt{-1}$.

A.3 SYNTHETIC DIVISION

In the example at the end of the previous section, the factor $x + 4$ is the divisor; $x^3 + 6x^2 + 13x + 20$ is the dividend; and $x^2 + 2x + 5$ is the quotient. The coefficient of x in the divisor is unity, and the coefficients of the leading terms just above the auxiliary total lines are 1, 2, and 5. These match the coefficients of the quotient. These facts may be incorporated into an array that displays the coefficients of both the dividend and the quotient, as well as the trial value of $b = -4$ in the diviser $x -$

$b = x + 4$:

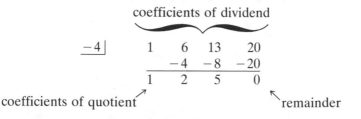

coefficients of dividend

$$\begin{array}{r|rrrr} -4 & 1 & 6 & 13 & 20 \\ & & -4 & -8 & -20 \\ \hline & 1 & 2 & 5 & 0 \end{array}$$

coefficients of quotient   remainder

Notice in the array that powers of x have been omitted because their position is clearly understood. Also notice that the leading coefficient of the quotient is unity as a consequence of the coefficient of x in $x + 4$ being equal to unity. Finally, notice that the second coefficient of the quotient comes from the subtraction of the product of the 1, which is the first coefficient in the quotient, and the 4, which is the constant term in the divisor. This is carried out in the array, but because a -4 is used, an addition rather than a subtraction is performed. This is handy because roots as well as factors are often sought.

The entire process is called *synthetic division* and is guided by five steps:

1. Write the coefficients of $f(x)$ in a horizontal line, putting in zeroes for the missing powers of x.
2. Assume a coefficient b and consider that $x - b$ is to be divided into $f(x)$. Write b at the left of the line of the coefficients of $f(x)$.
3. Leave space for a second line and write the first coefficient in the first line as the first or leading coefficient in the third line. Make sure that this coefficient is directly below its position in the first line.
4. The product of b and the coefficient in the third line are written beneath the second coefficient in the first line. This results in a column of two numbers that should be added and the sum placed in the third line.
5. Continue the process until the last term of the third line is reached. This term is the remainder, and if it is zero, b is a root of $f(x)$.

Example: Use synthetic division to find two real roots of

$$f(x) = x^4 + 7x^3 + 14x^2 + 14x + 24 = 0$$

Solution:

Descartes's rule of signs indicates that there are no positive real roots and that because

$$f(-x) = x^4 - 7x^3 + 14x^2 - 14x + 24 = 0$$

there can be as many as four negative real roots. Observe that $f(0) = 24$ so that the synthetic division may begin with $b = -1$.

$$
\begin{array}{r|rrrrr}
-1 & 1 & 7 & 14 & 14 & 24 \\
 & & -1 & -6 & -8 & -6 \\
\hline
 & 1 & 6 & 8 & 6 & 18 \\
-2 & 1 & 7 & 14 & 14 & 24 \\
 & & -2 & -10 & -8 & -12 \\
\hline
 & 1 & 5 & 4 & 6 & 12 \\
-3 & 1 & 7 & 14 & 14 & 24 \\
 & & -3 & -12 & -6 & -24 \\
\hline
-4 & 1 & 4 & 2 & 8 & 0 \\
 & & -4 & 0 & -8 & \\
\hline
 & 1 & 0 & 2 & 0 & \\
\end{array}
$$

Observe that -3 and -4 are roots and that the quotient or second depressed equation lacks a term involving x, $x^2 + 2 = 0$. Also observe that when the remainder is zero, which indicates a root, the process may continue without rewriting the array, because the depressed equation or quotient is the first line in the new array.

A.4 TRANSFORMATION OF POLYNOMIAL EQUATIONS

Consider the polynomial equation given in Eq. (A.1):

$$f(x) = a_n x^n + a_{n-1} x^{n-1} + a_{n-2} x^{n-2} + \cdots + a_1 x + a_0 = 0$$

that has n roots, $r_1, r_2, r_3, \cdots, r_{n-1}$ and r_n. In factored form, Eq. (A.2):

$$f(x) = a_n(x - r_1)(x - r_2)(x - r_3) \cdots (x - r_{n-1})(x - r_n)$$

Clearly, one may add or subtract a constant h to or from x so that $x = y + h$ or $y = x - h$. In this event, the roots of

$$f(y + h) = a_n(y + h - r_1)(y + h - r_2)$$
$$(y + h - r_3) \cdots (y + h - r_n) = 0 \tag{A.3}$$

can be obtained by setting any or all of the factors equal to zero. This yields $y = r_1 - h$, $y = r_2 - h$, $y = r_3 - h$, $\cdots$, $y = r_n - h$. The desired equation with each root reduced by h can be found from Eq. (A.1) by writing

$$f(y = h) = a_n(y + h)^n + a_{n-1}(y + h)^{n-1}$$
$$+ a_{n-2}(y + h)^{n-2} + \cdots + a_0 = 0$$

which can be expanded to give

$$f(y) = c_n y^n + c_{n-1} y^{n-1} + c_{n-2} y^{n-2} + \cdots + c_1 y + c_0 = 0 \quad \text{(A.4)}$$

where the c's do not necessarily equal the a's.

Because $x = y + h$ and $f(x) = f(y + h)$, which is a function of y, Eq. (A.4) can be written as

$$f(x) = c_n(x - h)^n + c_{n-1}(x - h)^{n-1} + c_{n-2}(x - h)^{n-2} + \cdots + c_0 = 0$$

This may be divided throughout by $x - h$:

$$\frac{f(x)}{x - h} = c_n(x - h)^{n-1} + c_{n-1}(x - h)^{n-2}$$
$$+ c_{n-2}(x - h)^{n-3} + \cdots + \frac{c_0}{x - h} = 0$$

This reveals that $c_0/(x - h)$ is the remainder R_0, which is the zeroth coefficient in the equation for $y = x - h$.

If the process of dividing by $x - h$ is continued n times, the equation

$$f(y) = a_n y^n + R_{n-1} y^{n-1} + R_{n-2} y^{n-2} + \cdots + R_1 y + R_0 = 0$$

is obtained.

Example: Consider

$$f(x) = x^3 + 9x^2 + 23x + 15 = 0$$

Find all the roots, develop an equation $f(y)$ where $y = x - 2$, and show that the roots of $f(y)$ are truly the roots of $f(x)$, each diminished by 2.

Solution:

Inspection of $f(x)$ shows that there are no positive roots and that because $n = 3$, there must be at least one negative root:

$$
\begin{array}{r|rrrr}
-1\underline{|} & 1 & 9 & 23 & 15 \\
 & & -1 & -8 & -15 \\
\hline
-2\underline{|} & 1 & 8 & 15 & 0 \\
 & & -2 & -12 & \\
\hline
 & 1 & 6 & 3 &
\end{array}
$$

which shows that -1 is a root of $f(x)$ and that -2 is not. To continue,

$$
\begin{array}{r|rrr}
-3| & 1 & 8 & 15 \\
 & & -3 & -15 \\ \hline
-4| & 1 & 5 & 0 \\
 & & -4 & \\ \hline
 & 1 & 1 &
\end{array}
$$

which shows that -3 is a root and that -4 is not. There is no need to proceed further, because the final depressed equation is $x + 5 = 0$, which yields $x = -5$ as the third root. Thus the roots of $f(x)$ are -1, -3, and -5.

An equation in $y = x - 2$ is found from the successive remainders in a synthetic division of $f(x)$ divided by $x + 2$:

$$
\begin{array}{r|rrrr}
2| & 1 & 9 & 23 & 15 \\
 & & 2 & 22 & 90 \\ \hline
2| & 1 & 11 & 45 & 105 \\
 & & 2 & 26 & \\ \hline
2| & 1 & 13 & 71 & \\
 & & 2 & & \\ \hline
 & 1 & 15 & &
\end{array}
$$

The numbers along the diagonal proceeding from upper right to lower left are the coefficients of $f(y)$ and they begin with at the upper right with R_0. Thus with $n = 3$ and $a_n = a_3 = 1$,

$$f(y) = y^3 + 15y^2 + 71y + 105 = 0$$

It is easy to calculate $f(0) = 105$ and $f(-1) = -1 + 15 - 71 + 105 = 48$. Thus the synthetic division may begin with $y = -2$ in this situation in which there can be no positive real roots and in which there must be at least one negative real root:

$$
\begin{array}{r|rrrr}
-2| & 1 & 15 & 71 & 105 \\
 & & -2 & -26 & -90 \\ \hline
 & 1 & 13 & 45 & 15 \\
-3| & 1 & 15 & 71 & 105 \\
 & & -3 & -36 & -105 \\ \hline
-4| & 1 & 12 & 35 & 0 \\
 & & -4 & -32 & \\ \hline
 & 1 & 8 & 3 &
\end{array}
$$

$$\begin{array}{r|rrr} -5\, \rfloor & 1 & 12 & 35 \\ & & -5 & -35 \\ \hline & 1 & 7 & 0 \end{array}$$

The foregoing shows that $y = -3$ and $y = -5$ are roots, and the final depressed equation $y + 7 = 0$ also indicates that $y = -7$ is a root. Thus the three roots of $f(y)$, -3, -5, and -7 which are each two less than the roots of $f(x)$, complete what the statement of the example requires.

A.5 HORNER'S METHOD

Horner's method permits a root-solving procedure by synthetic division using single-digit multipliers. Although this is not a major saving in the presence of a computer, it can be helpful in applications in which calculations are done by hand. It is based on depressing the roots of the given equation at the end of each calculation step.

Example: Find one real root of

$$f(x) = x^3 + 2x^2 + 5x + 8 = 0$$

and correct to the nearest thousandth.
Solution:
The equation must have one real root that cannot be positive. Observe that $f(0) = 8$ and begin with $x = -1$:

$$\begin{array}{r|rrrr} -1\,\rfloor & 1 & 2 & 5 & 8 \\ & & -1 & -1 & -4 \\ \hline & 1 & 1 & 4 & 4 \\ -2\,\rfloor & 1 & 2 & 5 & 8 \\ & & -2 & 0 & -10 \\ \hline & 1 & 0 & 5 & -2 \end{array}$$

This shows that a root lies between -1 and -2. An equation in $y = x + 1$ must now be found:

$$\begin{array}{r|rrrr} -1\,\rfloor & 1 & 2 & 5 & 8 \\ & & -1 & -1 & -4 \\ \hline -1\,\rfloor & 1 & 1 & 4 & 4 \\ & & -1 & 0 & \\ \hline -1\,\rfloor & 1 & 0 & 4 & \\ & & -1 & & \\ \hline & 1 & -1 & & \end{array}$$

and with $n = 3$ and $a_n = a_3 = 1$,

$$f(y) = y^3 - y^2 + 4y + 4 = 0$$

This equation must have a negative root between 0 and -1, because $f(0) = 4$ and $f(-1) = -1 -1 - 4 + 4 = -2$. This root of $f(y)$ must be exactly 1 more than the root of the original equation. The synthetic division should begin at about $y = -0.6$:

-0.6|	1	-1	4	4
			-0.6	0.96	-2.976
		1	-1.6	4.96	1.024
-0.7|	1	-1	4	4
			-0.7	1.19	-3.633
		1	-1.7	5.19	0.367
-0.8|	1	-1	4	4
			-0.8	1.44	-4.352
		1	-1.8	5.44	-0.352

A root of $f(y)$ lies between -0.7 and -0.8. An equation in $z = y + 0.7$ must now be found. This equation has a negative root that is exactly 0.7 more than a root of $f(y) = 0$ and a root that is exactly 1.7 more than a root of $f(x) = 0$:

-0.7|	1	-1	4	4
			-0.7	1.19	-3.633
-0.7|	1	-1.7	5.19	0.367
			-0.7	1.68	
-0.7|	1	-2.4	6.87	
			-0.7		
		1	-3.1		

Thus

$$f(z) = z^3 - 3.1z^2 + 6.87z + 0.367 = 0$$

Because the next step deals with values of z less than 0.1, the z^3 and z^2 terms of $f(z)$ can be neglected in finding an estimate for a root of $f(z)$. Thus

$$z = -\frac{0.367}{6.87} = -0.053$$

and one may begin with $z = -0.05$:

$$
\begin{array}{r|rrrr}
-0.05 & 1 & -3.1 & 6.87 & 0.367 \\
       &   & -0.05 & 0.158 & -0.351 \\
\hline
       & 1 & -3.15 & 7.028 & 0.016 \\
-0.06 & 1 & -3.1 & 6.87 & 0.367 \\
       &   & -0.06 & 0.190 & -0.424 \\
\hline
       & 1 & -3.16 & 7.060 & -0.057
\end{array}
$$

which shows that the root of $f(z)$ lies between -0.05 and -0.06. The next step is to form $f(w)$ with $w = z + 0.05$. Thus

$$
\begin{array}{r|rrrr}
-0.05 & 1 & -3.1 & 6.8700 & 0.367000 \\
       &   & -0.05 & 0.1575 & -0.351375 \\
\hline
-0.05 & 1 & -3.15 & 7.0275 & 0.015625 \\
       &   & -0.05 & 0.1600 & \\
\hline
-0.05 & 1 & -3.20 & 7.1875 & \\
       &   & -0.05 & & \\
\hline
       & 1 & -3.35 & &
\end{array}
$$

and

$$ f(w) = w^3 - 3.35w^2 + 7.1875w + 0.015625 = 0 $$

which has an approximate root at

$$ w = -\frac{0.015625}{7.1875} = -0.002 $$

Whether -0.002 yields a root of $f(x)$ to the nearest thousandth remains to be checked:

$$
\begin{array}{r|rrrr}
-0.002 & 1 & -3.25 & 7.187500 & 0.01562500 \\
        &   & -0.002 & 0.006504 & -0.01438801 \\
\hline
        & 1 & -3.252 & 7.194004 & 0.00123699 \\
-0.003 & 1 & -3.25 & 7.187500 & 0.1562500 \\
        &   & -0.003 & 0.009759 & -0.02159178 \\
\hline
        & 1 & -3.253 & 7.197259 & -0.00596678
\end{array}
$$

The root of $f(w)$ is closer to -0.002 than it is to -0.003. Thus the root of $f(x)$, to the closest thousandth, is $x = -1 -0.7 -0.05 -0.002$, or -1.752.

A.6 ALGEBRAIC SOLUTION OF QUADRATIC, CUBIC, AND QUARTIC EQUATIONS

A.6.1 Quadratic Equation

The quadratic equation

$$ax^2 + bx + c = 0 \tag{A.5}$$

can be solved by the "formula" that is subject to almost universal recall. With roots x_1 and x_2,

$$x_1, x_2 = \frac{-b \pm \sqrt{b^2 - 4ac}}{2a} \tag{A.6}$$

where $b^2 - 4ac$ is known as the *discriminant*.

It is often useful, particularly in applications involving the Laplace transformation, to solve the quadratic by a method known as completing the square. If b in Eq. (A.5) is divided by a, halved, then squared, and finally added to and subtracted from Eq. (A.5), the result is:

$$x^2 + \frac{b}{a}x + \left(\frac{b}{2a}\right)^2 + \frac{c}{a} - \left(\frac{b}{2a}\right)^2 = 0$$

The leading three terms form a perfect square, so that

$$\left(x + \frac{b}{2a}\right)^2 = \left(\frac{b}{2a}\right)^2 - \frac{c}{a}$$

and then a little algebra gives

$$x + \frac{b}{2a} = \pm \sqrt{\left(\frac{b}{2a}\right)^2 - \frac{c}{a}}$$

or

$$x_1, x_2 = -\frac{b}{2a} \pm \sqrt{\left(\frac{b}{2a}\right)^2 - \frac{c}{a}} \tag{A.7}$$

which can also be obtained by adjusting Eq. (A.6).

Example: Find the two roots of

$$x^2 + 4x + 5 = 0$$

by completing the square.

Solution:

$$x^2 + 4x + 4 + 5 - 4 = 0$$

$$(x + 2)^2 + 1 = 0$$

$$(x + 2)^2 = -1$$

$$x + 2 = \pm j$$

$$x_1, x_2 = -2 \pm j$$

By the quadratic formula of Eq. (A.6) with $a = 1$, $b = 4$, and $c = 5$,

$$x_1, x_2 = \frac{-4 \pm \sqrt{(-4)^2 - 4(1)(5)}}{2(1)}$$

$$= -2 \pm \frac{\sqrt{-4}}{2} = -2 \pm \frac{j2}{2}$$

or

$$x_1, x_2 = -2 \pm j$$

A.6.2 Cubic Equation

In the cubic equation:

$$f(x) = a_3 x^3 + a_2 x^2 + a_1 x + a_0 = 0$$

divide through by a_3 so that

$$f(x) = x^3 + bx^2 + cx + d = 0 \tag{A.8}$$

where $b = a_2/a_3$, $c = a_1/a_3$, and $d = a_0/a_3$. Then let

$$x = y - \frac{b}{3}$$

and substitute this into Eq. (A.8) to obtain after some algebraic simplification:

$$y^3 + py + q = 0 \tag{A.9}$$

an equation without a second-degree term that is called the *reduced cubic* where

$$p = c - \frac{b^2}{3} \tag{A.10a}$$

and

$$q = d - \frac{bc}{3} + \frac{2b^3}{27} \qquad \text{(A.10}b\text{)}$$

Now let a root of Eq. (A.9) be composed of a sum of two new variables u and v so that

$$y = u + v$$

and impose the additional requirement that

$$p + 3uv = 0 \qquad \text{(A.11}a\text{)}$$

If these are placed in Eq. (A.9), the result is

$$u^3 + v^3 + (p + 3uv)(u + v) + q = 0$$

or

$$u^3 + v^3 + q = 0 \qquad \text{(A.11}b\text{)}$$

Elimination of v between Eqs. (A.11) results in

$$u^3 - \frac{p^3}{27u^3} + q = 0$$

that is a quadratic in u^3:

$$27u^6 + 27qu^3 - p^3 = 0$$

that has two solutions:

$$u^3 = \frac{-27q \pm \sqrt{729q^2 + 108p^3}}{54}$$

or

$$u^3 = -\frac{q}{2} \pm \sqrt{R}$$

where

$$R = \frac{q^2}{4} + \frac{p^3}{27} \qquad \text{(A.12)}$$

Let

$$u^3 = -\frac{q}{2} + \sqrt{R} = A \qquad \text{(A.13}a\text{)}$$

so that from Eq. (A.11b).

$$v^3 = -\frac{q}{2} - \sqrt{R} = B \qquad \text{(A.13}b\text{)}$$

Observe that if u^3 were selected $-q/2 + \sqrt{R}$, then v^3 would have been $-q/2 - \sqrt{R}$ which is merely an interchange of u and v.

Any number N must have three cube roots, and these must be spaced $120°$ apart in the complex plane. If $\sqrt[3]{N}$ is one of them, then it is seen that the cubics $u^3 = A$ and $v^3 = B$ have the following roots:

$$u_1 = \sqrt[3]{A} \qquad\qquad v_1 = \sqrt[3]{B}$$

$$u_2 = \omega\sqrt[3]{A} \qquad\qquad v_2 = \omega\sqrt[3]{B}$$

$$u_3 = \omega^2 \sqrt[3]{A} \qquad\qquad v_3 = \omega^2 \sqrt[3]{B}$$

where

$$\omega = -\frac{1}{2} + j\frac{\sqrt{3}}{2}$$

The only pairs of roots of u and v that satisfy Eq. (A.11a) are:

$$y_1 = \sqrt[3]{A} + \sqrt[3]{B}$$

$$y_2 = \omega\sqrt[3]{A} + \omega^2 \sqrt[3]{B}$$

and

$$y_3 = \omega^2 \sqrt[3]{A} + \omega\sqrt[3]{B}$$

and if these are used in $x = y - b/3$, then the three roots of the cubic equation $f(x) = 0$ are:

$$x_1 = y_1 - \frac{b}{3}$$

$$x_2 = y_2 - \frac{b}{3}$$

and

$$x_3 = y_3 - \frac{b}{3}$$

Example: Find the three roots of the following cubic equation:

$$f(x) = x^3 + 3x^2 + 7x + 5 = 0$$

Solution:
Here $b = 3$, $c = 7$, and $d = 5$ so that by Eqs. (A.10),

$$p = c - \frac{b^2}{3} = 7 - \frac{(3)^2}{3} = 7 - 3 = 4$$

and

$$q = d - \frac{bc}{3} + \frac{2b^3}{27} = 5 - \frac{3(7)}{3} + \frac{2(3)^3}{27} = 5 - 7 + 2 = 0$$

These values of p and q are used in Eq. (A.12) to find R:

$$R = \frac{q^2}{4} + \frac{p^3}{27} = 0 + \frac{(4)^3}{27} = \frac{64}{27}$$

so that from Eqs. (A.13),

$$A = -\frac{q}{2} + \sqrt{R} = 0 + \sqrt{\frac{64}{27}} = \frac{8\sqrt{3}}{9}$$

and because $q = 0$

$$B = -A = -\frac{8\sqrt{3}}{9}$$

Then with $\omega^2 = -\frac{1}{2} - j\frac{\sqrt{3}}{2}$,

$$y_1 = \sqrt[3]{A} + \sqrt[3]{B} = \sqrt[6]{\frac{64}{27}} - \sqrt[6]{\frac{64}{27}} = 0$$

$$y_2 = \omega\sqrt[3]{A} + \omega^2 \sqrt[3]{B} = \left(-\frac{1}{2} + j\frac{3}{2}\right)\sqrt[3]{A} + \left(-\frac{1}{2} - j\frac{3}{2}\right)\sqrt[3]{B} = +j\sqrt{3}\,\sqrt[3]{A}$$

and

$$y_3 = \omega^2 \sqrt[3]{A} + \omega\sqrt[3]{A} = \left(-\frac{1}{2} - j\frac{3}{2}\right)\sqrt[3]{A} + \left(-\frac{1}{2} + j\frac{3}{2}\right)\sqrt[3]{A} = -j\sqrt{3}\,\sqrt[3]{A}$$

The roots are, therefore,

$$x_1 = y_1 - \frac{b}{3} = 0 - \frac{3}{3} = -1$$

$$x_2 = y_2 - \frac{b}{3} = j\sqrt{3}\,\sqrt[3]{A} - \frac{3}{3} = -1 + j\sqrt{3}\,\sqrt[6]{\frac{64}{27}} = -1 + j2$$

and

$$x_3 = y_3 - \frac{b}{3} = -j\sqrt{3}\,\sqrt[3]{A} - \frac{3}{3} = -1 - j\sqrt{3}\,\sqrt[6]{\frac{64}{27}} = -1 - j2$$

A.6.3 Quartic or Biquadratic Equation

The quartic or biquadratic equation:

$$f(x) = a_4x^4 + a_3x^3 + a_2x^2 + a_1x + a_0 = 0$$

can be divided through by a_4 so that

$$f(x) = x^4 + bx^3 + cx^2 + dx + e = 0 \qquad \text{(A.14)}$$

where $b = a_3/a_4$, $c = a_2/a_4$, $d = a_1/a_4$, and $e = a_0/a_4$ and Eq. (A.14) can be written as the difference of two squares:

$$(x^2 + Ax + B)^2 - (Cx + D)^2 = 0$$

or

$$x^4 + 2Ax^3 + (A^2 + 2B - C^2)x^2$$
$$+ (2AB - 2CD)x + (B^2 - D^2) = 0 \qquad \text{(A.15)}$$

If in Eqs. (A.14) and (A.15) the coefficients of like powers of x are equated, a set of four linear simultaneous algebraic equations are obtained:

$$2A = b \qquad \text{(A.16a)}$$

$$A^2 + 2B - C^2 = c \qquad \text{(A.16b)}$$

$$2AB - 2CD = d \qquad \text{(A.16c)}$$

$$B^2 - D^2 = e \qquad \text{(A.16d)}$$

Elimination of A, C, and D in Eqs. (A.16) shows that $2B$ is a root of the cubic equation

$$y^3 - cy^2 + fy + g = 0 \qquad \text{(A.17)}$$

which is called the *resolvent cubic* where

$$f = bd - 4e \qquad \text{(A.18a)}$$

and

$$g = e(4c - b^2) - d^2 \qquad \text{(A.18b)}$$

Now because $2B$ is a root of Eq. (A.17),

$$B = \frac{y_1}{2} \qquad \text{(A.19a)}$$

where y_1 is taken as the algebraically largest real root of Eq. (A.17). Equations (A.16) can then be solved:

$$A = \frac{b}{2} \qquad \text{(A.19b)}$$

$$D = \sqrt{B^2 - e} \qquad \text{(A.19c)}$$

and

$$C = \frac{AB - d/2}{D} \qquad (D \neq 0) \qquad \text{(A.19d)}$$

or

$$C = \sqrt{A^2 + 2B} - C \qquad (D = 0) \qquad \text{(A.19e)}$$

Use of these values in Eq. (A.15) yields two quadratic equations:

$$x^2 + (A - C)x + (B - D) = 0 \qquad \text{(A.20a)}$$

and

$$x^2 + (A + C)x + (B + D) = 0 \qquad \text{(A.20b)}$$

each of which can be solved for two of the roots of Eq. (A.14).

Example: The quartic equation

$$f(x) = x^4 + 2x^3 + 5x^2 + 4x + 3 = 0$$

has no real roots. Find all the roots.
Solution:
Here, $b = 2$, $c = 5$, $d = 4$, and $e = 3$. By Eqs. (A.18),

$$f = bd - 4e = 2(4) - 4(3) = 8 - 12 = -4$$

and

$$g = e(4c - b^2) - d^2 = 3[4(5) - 2^2] - 4^2 = 3(16) - 16 = 32$$

The resolvent cubic equation is given by Eq. (A.17):

$$y^3 - 5y^2 - 4y + 32 = 0$$

and it can be shown by synthetic division that one root is $y_1 = 4$:

$$
\begin{array}{r|rrrr}
4 & 1 & -5 & -4 & 32 \\
  &   & 4 & -4 & -32 \\
\hline
  & 1 & -1 & -8 & 0 \\
\end{array}
$$

and that the other two roots are the roots of the quadratic:

$$y^2 - y - 8 = 0$$

By the quadratic formula of Eq. (A.6), $y_2 = 3.3723$ and $y_3 = -2.3723$

and hence $y_1 = 4$ is the largest algebraic root. Thus with $y_1 = 4$, by Eqs. (A.19),

$$B = \frac{y_1}{2} = \frac{4}{2} = 2$$

$$A = \frac{b}{2} = \frac{2}{2} = 1$$

$$D = \sqrt{b^2 - e} = \sqrt{(2)^2 - 3} = 1$$

and because $D \neq 0$,

$$C = \frac{AB - \dfrac{d}{2}}{D} = \frac{2(1) - \frac{1}{2}(4)}{1} = 0$$

The two quadratics to be solved are obtained from Eqs. (A.20) and with $C = 0$, $A = D = 1$, and $B = 2$, these are:

$$x^2 + (A - C)x + (B - D) = x^2 + x + 1 = 0$$

and

$$x^2 + (A + C)x + (B + D) = x^2 + x + 3 = 0$$

The roots of the first are:

$$x_1, x_2 = -\frac{1}{2} \pm j\frac{3}{2}$$

and the roots of the second are

$$x_3, x_4 = -\frac{1}{2} \pm j\frac{11}{2}$$

A.7 GRAEFFE'S ROOT SQUARING METHOD

A.7.1 Introduction

Greaffe's root squaring method can find multiple complex conjugate roots of a polynomial equation. It is based on the formation of a sequence of polynomial equations of the same degree as the original equation. The last equation in the sequence has roots that are some large even power of the roots of the original equation, and by certain relationships that exist between the coefficients of the last equation, the roots of the original equation may be extracted by a logarithmic process.

Consider the third-order cubic equation ($n = 3$):

$$f(x) = x^3 + a_1x^2 + a_2x + a_3 = 0 \qquad (A.21)$$

that has roots x_1, x_2, and x_3 and the polynomial equation:

$$f(-x) = -x^3 + a_1x^2 - a_2x + a_3 = 0$$

that has roots $-x_1$, $-x_2$, and $-x_3$. Multiplication of the two gives

$$f(x)f(-x) = -x^6 + (a_1^2 - 2a_2)x^4 - (a_2^2 - 2a_1a_3)x^2 + a_3^2 = 0$$

If $y = -x^2$, then

$$y^3 + (a_1^2 - 2a_2)y^2 + (a_2^2 - 2a_1a_3)y + a_3^3 = 0$$

or

$$y^3 + b_1y^2 + b_2y + b_3 = 0 \qquad (A.22)$$

where

$$b_1 = a_1^2 - 2a_2(1) \qquad (A.23a)$$

$$b_2 = a_2^2 - 2a_1a_3 \qquad (A.23b)$$

and

$$b_3 = a_3^2 - 2a_2(0) \qquad (A.23c)$$

Equation (A.22) is called the *derived polynomial* and is of the same degree as the original polynomial of Eq. (A.21). Each of the roots of the derived polynomial are the negative of the squares of the roots of the original polynomial. Another derived polynomial can be obtained if the procedure is repeated, treating the first derived polynomial of Eq. (A.22) as the original polynomial.

Consider Eqs. (A.23), which provide the coefficients for the derived polynomial in terms of the coefficients of the original polynomial. Each of these coefficients involve two terms; the first term is the *square of the coefficient* of the original polynomial, and the second is called the *doubled product*. The double product consists of the product of the coefficients on either side of the coefficient in the original polynomial multipled by two.

For a general nth-order polynomial of the form

$$f(x) = x^n + a_1x^{n-1} + a_2x^{n-2} + \cdots + a_{n-1}x + a_n = 0$$

the derived polynomial is:

$$f(y) = y^n + b_1y^{n-1} + b_2y^{n-2} + \cdots + b_{n-1}y + b_n = 0$$

where

$$b_1 = a_1^2 - 2a_2(1) = a_1^2 - 2a_2$$

$$b_2 = a_2^2 - 2a_1a_3 + 2a_4(1) = a_2^2 - 2a_1a_3 + 2a_4$$

$$b_3 = a_3^2 - 2a_2a_4 + 2a_1a_5 + 2a_6$$

$$\cdots$$

$$b_{n-1} = a_{n-1}^2 - 2a_{n-2}a_n$$

$$b_n = a_n^2 - 2a_{n-1}(0) = a_n^2$$

Notice that the doubled products alternate in sign.

The root squaring process may be conveniently tabulated. Each step in the procedure can be expeditiously represented by a numeral $m = 1$, $2, 4, 8, 16, 32, \ldots$ so that the final derived equation can inherently include the number of steps involved in its formation. The root squaring process may be terminated when the coefficients of a derived equation are approximately the squares of the like coefficients in the derived equation immediately preceding.

Example: Take

$$f(x) = x^4 + 10x^3 + 35x^2 + 50x + 24 = 0$$

and find the derived equation for $m = 32$.

Solution:

The solution as determined from Table A.1 is:

$$y^4 + 1.844860 \times 10^{19}y^3 + 3.418227 \times 10^{34}y^2$$
$$+ 1.468114 \times 10^{44}y + 1.468114 \times 10^{44} = 0$$

A.7.2 All Roots Real and Distinct

If it is assumed that the roots of the original equation

$$f(x) = x^n + a_1x^{n-1} + a_2x^{n-2} + \cdots + a_{n-1}x + a_0 = 0$$

$x_1, x_2, x_3, \ldots, x_n$ are of such magnitude that

$$|x_1| > |x_2| > |x_3| > \cdots > |x_n|$$

then the root squaring procedure is concluded when

$$|x_1^m| \gg |x_2^m| \gg |x_3^m| \gg \cdots \gg |x_n^m| \qquad \text{(A.24)}$$

Table A.1 Root squaring tabulation ($n = 4$)

Numbers in parentheses are powers of 10

m		1^2	$a_1^2 - 2a_2$	$a_2^2 - 2a_1a_3 + 2a_4$	$a_3^2 - 2a_2a_4$	a_4^2
		1	10	35	50	24
1	Squared Coefficients	1	1.000000 (2)	1.225000 (3)	2.500000 (3)	5.760000 (2)
	Doubled Products		−7.000000 (1)	−1.000000 (3)	−1.680000 (3)	
				4.800000 (1)		
2	Squared Coefficients	1	3.000000 (1)	2.730000 (2)	8.200000 (2)	5.760000 (2)
	Doubled Products		9.000000 (2)	7.452900 (4)	6.724000 (5)	3.317760 (5)
			−5.460000 (2)	−4.920000 (4)	−3.144960 (5)	
				1.152000 (3)		
4	Squared Coefficients	1	3.540000 (2)	2.648100 (4)	3.579040 (5)	3.317760 (5)
	Doubled Products		1.253160 (5)	7.012434 (8)	1.280953 (11)	1.100753 (11)
			−5.296200 (4)	−2.533960 (8)	−1.757152 (10)	
				6.635520 (5)		
8	Squared Coefficients	1	7.235400 (4)	4.485109 (8)	1.105238 (11)	1.100753 (11)
	Doubled Products		5.235101 (9)	2.011620 (17)	1.221550 (22)	1.211657 (22)
			−8.970218 (8)	−1.599367 (16)	−9.873995 (19)	
				2.201506 (11)		
16	Squared Coefficients	1	4.338080 (9)	1.851686 (17)	1.211676 (22)	1.211657 (22)
	Doubled Products		1.881893 (19)	3.428740 (34)	1.468159 (44)	1.468114 (44)
			−3.703371 (17)	−1.051269 (32)	−4.487217 (39)	
				2.423315 (22)		
32		1	1.844860 (19)	3.418227 (34)	1.468114 (44)	1.468114 (44)

in the nth-degree derived equation

$$f(y) = y^n + b_1 y^{n-1} + b_2 y^{n-2} + \cdots + b_{n-1} y + b_n = 0 \quad \text{(A.25)}$$

with roots $y = -x^{2m}$.

   If a set of roots that are negatives of the actual roots is introduced

$$z_1 = -x_1$$

$$z_2 = -x_2$$

$$z_3 = -x_3$$

$$\cdots$$

$$z_n = -x_n$$

then Eq. (A.25) which can be written in factored form:

$$(y - y_1)(y - y_2)(y - y_3) \cdots (y - y_n) = 0$$

can also be written as

$$(y + z_1^m)(y + z_2^m)(y + z_3^m) \cdots (y + z_n^m) = 0 \quad \text{(A.26)}$$

Expansion of Eq. (A.26) gives:

$$y^n + (z_1^m + z_2^m + z_3^m + \cdots + z_n^m) y^{n-1}$$
$$+ (z_1^m z_2^m + z_1^m z_3^m + \cdots + z_2^m z_3^m + z_2^m z_4^m + \cdots) y^{n-2}$$
$$+ (z_1^m z_2^m z_3^m + z_1^m z_2^m z_4^m + z_1^m z_3^m z_4^m + \cdots) y^{n-3}$$
$$+ \cdots + (z_1^m z_2^m z_3^m z_4^m z_5^m \cdots z_n^m) = 0 \quad \text{(A.27)}$$

Clearly, Eqs. (A.26) and (A.27) reveal that the roots of Eq. (A.26) can be related to the coefficients of Eq. (A.25) by

$$b_1 = z_1^m + z_2^m + z_3^m + \cdots + z_n^m \quad \text{(A.28a)}$$

$$b_2 = z_1^m z_2^m + z_1^m z_3^m + \cdots + z_2^m z_3^m + z_2^m z_4^m + \cdots \quad \text{(A.28b)}$$

$$b_3 = z_1^m z_2^m z_3^m + z_1^m z_2^m z_4^m + z_1^m z_3^m z_4^m + \cdots \quad \text{(A.28c)}$$

$$\cdots$$

$$b_n = z_1^m z_2^m z_3^m z_4^m \cdots z_n^m \quad \text{(A.28d)}$$

Under the conditions of Eq. (A.24) or

$$|z_1^m| \gg |z_2^m| \gg |z_3^m| \gg \cdots \gg |z_n^m| \quad \text{(A.29)}$$

the coefficients b_k in Eqs. (A.18) can be approximated by

$$b_1 \approx z_1^m$$

$$b_2 \approx z_1^m z_2^m$$

$$b_3 \approx z_1^m z_2^m z_3^m$$

$$\cdots$$

$$b_n \approx z_1^m z_2^m z_3^m \cdots z_n^m$$

and these equations provide the roots z_1 to z_n:

$$z_1 = \pm (b_1)^{1/m} \qquad\qquad (A.30a)$$

$$z_2 = \pm \left(\frac{b_2}{b_1}\right)^{1/m} \qquad\qquad (A.30b)$$

$$z_3 = \pm \left(\frac{b_3}{b_2}\right)^{1/m} \qquad\qquad (A.30c)$$

$$\cdots$$

$$z_n = \pm \left(\frac{b_n}{b_{n-1}}\right)^{1/m} \qquad\qquad (A.30d)$$

The determination of whether the plus or minus should be used can come from an actual substitution of the values of z_1 through z_n into the original polynomial equation or a use of Descartes's rule of signs.

Example: The polynomial equation

$$x^4 + 10x^3 + 35x^2 + 50x + 24 = 0$$

has, as the reader may verify, roots of -1, -2, -3, and -4. The example in the previous section showed that, with $m = 32$, the coefficients of the derived equation are:

$$b_1 = 1.844860 \times 10^{19}$$

$$b_2 = 3.418227 \times 10^{34}$$

$$b_3 = 1.468114 \times 10^{44}$$

and

$$b_4 = 1.468114 \times 10^{44}$$

Verify that -1, -2, -3, and -4 are the roots.

Solution:

By Eqs.(A.30):

$$z_1 = \pm (b_1)^{1/m} = \pm (1.844860 \times 10^{19})^{1/32} = \pm 4.000013$$

$$z_2 = \pm \left(\frac{b_2}{b_1}\right)^{1/m} = \pm \left(\frac{3.418227 \times 10^{34}}{1.844860 \times 10^{19}}\right)^{1/32} = \pm (1.852838 \times 10^{15})^{1/32}$$

or

$$z_2 = \pm 2.999991$$

$$z_3 = \pm \left(\frac{b_3}{b_2}\right)^{1/m} = \pm \left(\frac{1.468114 \times 10^{44}}{3.418227 \times 10^{34}}\right)^{1/32} = \pm (4.294958 \times 10^{9})^{1/32}$$

or

$$z_3 = \pm 2.000000$$

and

$$z_4 = \pm \left(\frac{b_4}{b_3}\right)^{1/m} = \pm \left(\frac{1.468114 \times 10^{44}}{1.468114 \times 10^{44}}\right)^{1/32} = \pm (1)^{1/32} = \pm 1.000000$$

Observe that the roots determined from the root squaring process are almost exactly equal to the actual roots and that the accuracy is quite sufficient for engineering computations. In this case, Descartes's rule of signs indicates that no roots can be positive. Thus the minus signs are selected, and -1.0000, -2.0000, -3.0000, and -4.0000 are the roots to the nearest ten-thousandth.

A.7.3 All Roots Real But Some Roots Repeated

If the roots of the polynomial equation are not all distinct, as m becomes large one or more of the coefficients of the derived equation will not approach the square of the immediately preceding coefficient. In fact, if the kth- and the $k - 1$th-roots are numerically equal, in general, the coefficient of y^{n-k} will only approach one-half of the square of the immediately preceding coefficient.

Again consider

$$f(x) = x^n + a_1 x^{n-1} + a_2 x^{n-2} + a_{n-1} x + a_n = 0$$

and suppose that $x_1, x_2, x_3, x_4, \ldots, x_n$ are the roots, with $x_3 = x_4$. Under these conditions, with $z = -x$, the derived equation is still Eq. (A.25) with $y = -x^{2m} = z^{2m}$, but the conditions of Eqs. (A.28) must be modified:

$$b_1 = z_1^m + z_2^m + 2z_3^m + \cdots + z_n^m \tag{A.31a}$$

$$b_2 = z_1^m z_2^m + 2z_1^m z_3^m + 2z_2 z_3^m + \cdots \tag{A.31b}$$

$$b_3 = 2z_1^m z_2^m z_3^m + z_1^m z_3^{2m} + \cdots \tag{A.31c}$$

$$b_4 = z_1^m z_2^m z_2^{2m} + \cdots \tag{A.31d}$$

$$\cdots$$

$$b_n = z_1^m z_2^m z_3^{2m} z_5^m \cdots z_n^m \tag{A.31e}$$

Under the conditions of Eq. (A.29), Eqs. (A.31) can be approximated by

$$b_1 \approx z_1^m$$

$$b_2 \approx z_1^m z_2^m$$

and because the terms in Eq. (A.31c) may be comparable in magnitude, b_3 is not considered:

$$b_4 \approx z_1^m z_2^m z_3^{2m}$$

$$\cdots$$

$$b_n \approx z_1^m z_2^m z_3^{2m} z_5^m \cdots z_n^m$$

and these equations provide the roots z_1 to z_n:

$$z_1 = \pm(b_1)^{1/m} \tag{A.32a}$$

$$z_2 = \pm\left(\frac{b_2}{b_1}\right)^{1/m} \tag{A.32b}$$

$$z_3 = z_4 = \pm\left(\frac{b_4}{b_2}\right)^{1/2m} \tag{A.32c}$$

$$z_5 = \pm\left(\frac{b_5}{b_4}\right)^{1/m} \tag{A.32d}$$

$$\cdots$$

$$z_n = \pm\left(\frac{b_n}{b_{n-1}}\right)^{1/m}$$

Example: The polynomial equation

$$f(x) = x^4 + 8x^3 + 23x^2 + 28x + 12 = 0$$

has roots -1, -2, -2, and -3. Verify this by the Graeffe root squaring method.

Solution:

The root squaring tabulation is displayed in Table A.2 and is carried out to $m = 64$ to achieve the greatest possible accuracy. The tabulation indicates that in all columns but the second, the coefficients do not approach the square of the previous value. In the second column, the coefficient approaches approximately the square of the previous value. Because this is a fourth-degree polynomial equation ($n = 4$) and this behavior is observed in the second or y^2 column, then $n - k = 4 - z = 2$, which indicates that the second ($k = 2$) and the third ($k + 1 = 3$) roots are equal.

The coefficients of the derived equation with $m = 64$ are:

$$b_1 = 3.433684 \times 10^{30}$$

$$b_2 = 1.266806 \times 10^{50}$$

$$b_3 = 1.168422 \times 10^{69}$$

$$b_4 = 1.168422 \times 10^{69}$$

and the roots z_1 to z_4 are:

$$z_1 = \pm(b_1)^{1/m} = \pm(3.433684 \times 10^{30})^{1/64} = \pm 3.000000$$

$$z_z = z_3 = \pm\left(\frac{b_3}{b_1}\right)^{1/2m} = \pm\left(\frac{1.168422 \times 10^{69}}{3.433684 \times 10^{30}}\right)^{1/128} = \pm(3.402823 \times 10^{38})^{1/128}$$

or

$$z_2 = z_3 = \pm 2.000000$$

and

$$z_4 = \pm\left(\frac{b_4}{b_3}\right)^{1/64} = \pm\left(\frac{1.168422 \times 10^{69}}{1.168422 \times 10^{69}}\right)^{1/64} = \pm(1)^{1/64} = \pm 1.000000$$

Because the polynomial equation can have no positive roots, the roots to the nearest ten-thousandth are -1.0000, -2.0000, -2.0000, and -3.0000.

A.7.4 Some Roots Real and Distinct and Some Roots Complex

If some of the roots of a polynomial equation are complex conjugates, a fluctuation in sign appears in the root squaring tabulation. In fact, if the kth and the $k + 1$th roots are complex conjugates, the coefficient of y^{n-k} fluctuates in sign.

Table A.2 Root squaring tabulation ($n = 4$)

Numbers in parentheses are powers of 10

m		1^2	$a_1^2 - 2a_2$	$a_2^2 - 2a_1a_3 + 2a_4$	$a_3^2 - 2a_2a_4$	a_4^2
		1	8	23	28	12
1	Squared Coefficients	1	6.400000 (1)	5.290000 (2)	7.840000 (2)	1.440000 (2)
	Doubled Products		−4.600000 (1)	−4.480000 (2)	−5.520000 (2)	
				2.400000 (1)		
2	Squared Coefficients	1	1.800000 (1)	1.050000 (2)	2.320000 (2)	1.440000 (2)
	Doubled Products		3.240000 (2)	1.102500 (4)	5.382400 (4)	2.073600 (4)
			−2.100000 (2)	−8.352000 (3)	−3.024000 (3)	
				2.880000 (2)		
4	Squared Coefficients	1	1.140000 (2)	2.961000 (3)	2.358400 (4)	2.073600 (4)
	Doubled Products		1.299600 (4)	8.767521 (6)	5.562051 (8)	4.299817 (8)
			−5.922000 (3)	−5.377152 (6)	−1.227986 (8)	
				4.147200 (4)		
8	Squared Coefficients	1	7.074000 (3)	3.431841 (6)	4.334065 (8)	4.299817 (8)
	Doubled Products		5.004148 (7)	1.177753 (13)	1.878412 (17)	1.848843 (17)
			−6.863682 (6)	−6.131835 (12)	−2.951258 (15)	
				8.599634 (8)		
16	Squared Coefficients	1	4.317779 (7)	5.646558 (12)	1.848899 (17)	1.848843 (17)
	Doubled Products		1.864322 (15)	3.188362 (25)	3.418428 (34)	3.418219 (34)
			−1.129312 (13)	−1.596628 (25)	−2.087919 (30)	
				3.697685 (17)		
32	Squared Coefficients	1	1.853029 (15)	1.591734 (25)	3.418219 (34)	3.418219 (34)
	Doubled Products		3.433716 (30)	2.533617 (50)	1.168422 (69)	1.168422 (69)
			−3.183468 (25)	−1.266812 (50)	−1.088179 (60)	
				6.836430 (34)		
64		1	3.433684 (30)	1.266806 (50)	1.168422 (69)	1.168422 (69)

Suppose that the roots are $x_1, x_2, x_3, x_4, \ldots, x_n$, where x_3 and x_4 are complex conjugates and where, as in previous considerations,

$$|x_1| > |x_2| > |x_3| > |x_4| > \cdots > |x_n|$$

Let $x_3 = \rho\angle\theta$ so that $x_4 = \rho\angle-\theta$, and with $z = -x$

$$z_3 = u + jv = \rho\angle\theta = \rho e^{j\theta}$$

and

$$z_4 = u - jv = \rho\angle-\theta = \rho e^{-j\theta}$$

The coefficient relationships of Eqs. (A.28) then become

$$b_1 = z_1^m + z_2^m + \rho^m(e^{jm\theta} + e^{-jm\theta}) + \cdots + z_n^m$$

or with $\cos\theta = \frac{1}{2}(e^{j\theta} + e^{-j\theta})$,

$$b_1 = z_1^m + z_2^m + 2z_3^m\rho^m \cos m\theta + \ldots + z_n^m \quad (A.33a)$$

$$b_2 = z_1^m z_2^m + 2z_1^m\rho^m \cos m\theta + 2z_2^m\rho^m \cos m\theta + \cdots$$
$$(A.33b)$$

$$b_3 = 2z_1^m z_2^m\rho^m \cos m\theta + 2z_2 z_5^m\rho^m \cos m\theta + \cdots \quad (A.33c)$$

$$b_4 = z^{\cdot m} z^m\rho^{2m} + z^m z^m\rho^{2m} + \cdots \quad (A.33d)$$

because $z_3^m z_4^m = \rho^{2m}$,

$$\ldots$$

$$b_n = z_1^m z_2^m\rho^{2m} z_5^m \cdots z_n^m \quad (A.33e)$$

If the z's are arranged consistent with Eq. (A.24),

$$|z_1^m| \gg |z_2^m| \gg |\rho^m| \gg \cdots \gg |z_n^m|$$

then Eqs. (A.33) can be approximated by:

$$b_1 \approx z_1^m$$

$$b_2 \approx z_1^m z_2^m$$

$$b_3 \approx 2z_1^m z_2^m\rho^m \cos m\theta$$

$$b_4 \approx z_1^m z_2^m\rho^{2m}$$

$$\ldots$$

$$b_n \approx z_1^m z_2^m\rho^{2m} \cdots z_n^m$$

This indicates that the coefficient b_3 fluctuates in sign because of the presence of the cosine term. This fluctuation reveals the presence of a pair of complex conjugate roots. The signs, of course, may not fluctuate in any regular pattern, because successive values of $m\theta$ are not necessarily separated by π radians.

The approximations for the b's permit the evaluation of all roots but the complex conjugate pair:

$$z_1 = \pm(b_1)^{1/m} \tag{A.34a}$$

$$z_2 = \pm\left(\frac{b_2}{b_1}\right)^{1/m} \tag{A.34b}$$

$$\cdots$$

$$z_n = \pm\left(\frac{b_n}{b_{n-1}}\right)^{1/m} \tag{A.34c}$$

The magnitude of the complex roots is obtained from b_4:

$$\rho = \left(\frac{b_4}{b_2}\right)^{1/2m} \tag{A.34d}$$

and, as the alert reader has surmised, the real and imaginary parts of the complex roots are determined, in part, from this magnitude.

It has been observed that in

$$f(x) = x^n + a_1x^{n-1} + a_2x^{n-2} + \cdots + a_{n-1}x + a_n = 0$$

the coefficient a_1 is the sum of all the roots. If u is the real part of both x_3 and x_4, which are complex conjugates, then

$$a_1 = x_1 + x_2 + 2u + x_5 + \cdots + x_n$$

and hence

$$u = \pm\tfrac{1}{2}(a_1 - x_1 - x_2 - x_3 - x_5 \cdots - x_n) \tag{A.35a}$$

With u so established, v, the imaginary part of the conjugate pair, is:

$$v = \sqrt{\rho^2 - u^2} \tag{A.35b}$$

Example: Find the roots of

$$f(x) = x^4 + 7x^3 + 18x^2 + 22x + 12 = 0$$

Solution:

The root squaring tabulation is shown in Table A.3 and is carried to $m = 64$. The fluctuation in sign in the third column indicates the presence of a pair of complex conjugate roots whose magnitude is less than the absolute value of either of the real roots. The coefficients of the derived equation are:

$$b_1 = 3.433684 \times 10^{30}$$

$$b_2 = 6.334029 \times 10^{49}$$

$$b_3 = 5.440889 \times 10^{59}$$

Table A.3 Root squaring tabulation ($n=4$)

Numbers in parentheses are powers of 10

m		1^2	$a_1^2 - 2a_2$	$a_2^2 - 2a_1a_3 + 2a_4$	$a_3^2 - 2a_2a_4$	a_4^2
		1	7	18	22	12
1	Squared Coefficients	1	4.900000 (1)	3.240000 (2)	4.840000 (2)	1.440000 (2)
	Doubled Products		−3.600000 (1)	−3.080000 (2)	−4.320000 (2)	
				2.400000 (1)		
2		1	1.300000 (1)	4.000000 (1)	5.200000 (1)	1.440000 (2)
	Squared Coefficients		1.690000 (2)	1.600000 (3)	2.704000 (3)	2.073600 (4)
	Doubled Products		−8.000000 (1)	−1.352000 (3)	−1.152000 (4)	
				2.880000 (2)		
4		1	8.900000 (1)	5.360000 (2)	−8.816000 (3)	2.073600 (4)
	Squared Coefficients		7.921000 (3)	2.872960 (5)	7.772186 (7)	4.299817 (8)
	Doubled Products		−1.072000 (3)	1.569248 (6)	−2.222899 (7)	
				4.147200 (4)		
8		1	6.849000 (3)	1.898016 (6)	5.549286 (7)	4.299817 (8)
	Squared Coefficients		4.690880 (7)	3.602465 (12)	3.079458 (15)	1.848843 (17)
	Doubled Products		−3.796032 (6)	−7.601413 (11)	−1.632224 (15)	
				8.599634 (8)		
16		1	4.311277 (7)	2.843183 (12)	1.447234 (15)	1.848843 (17)
	Squared Coefficients		1.858711 (15)	8.083692 (24)	2.094485 (30)	3.418219 (34)
	Doubled Products		−5.686367 (12)	−1.247885 (23)	−1.051320 (30)	
				3.697685 (17)		
32		1	1.853024 (15)	7.958904 (24)	1.043166 (30)	3.418219 (34)
	Squared Coefficients		3.433700 (30)	6.334415 (49)	1.088194 (60)	1.168422 (69)
	Doubled Products		−1.591781 (25)	−3.866023 (45)	−5.441055 (59)	
				6.836438 (34)		
64		1	3.433684 (30)	6.334029 (49)	5.440889 (59)	1.168422 (69)

and

$$b_4 = 1.168422 \times 10^{69}$$

The real roots are established first, because the sign fluctuation in the third column of the root squaring tabulation indicates that the complex roots are the last two:

$$z_1 = \pm(b_1)^{1/m} = \pm(3.433684 \times 10^{30})^{1/64} = \pm 3.000000$$

and

$$z_2 = \pm\left(\frac{b_2}{b_1}\right)^{1/m} = \pm\left(\frac{6.334029 \times 10^{49}}{3.433684 \times 10^{30}}\right)^{1/64} = \pm(1.844674 \times 10^{19})^{1/64}$$

or

$$z_2 = \pm 2.000000$$

The minus sign is selected because $f(x) = 0$ cannot have a positive real root. Hence $x_1 = -3.0000$ and $x_2 = -2.0000$.

From Eq. (A.34d):

$$\rho = \left(\frac{b_4}{b_2}\right)^{1/2m} = \left(\frac{1.168422 \times 10^{69}}{6.334029 \times 10^{49}}\right)^{1/128} = (1.844674)^{1/128}$$

$$= 1.414212 = \sqrt{2}$$

and by Eq. (A.35a):

$$u = \pm\tfrac{1}{2}(a_1 - x_1 - x_2) = \pm\tfrac{1}{2}(7 - 3 - 2) = \pm\tfrac{1}{2}(2) = \pm 1$$

Then by Eq. (A.35b):

$$v = \sqrt{\rho^2 - u^2} = \sqrt{(\sqrt{2})^2 - (1)^2} = \sqrt{2 - 1} = 1 = 1$$

This makes

$$x_3, x_4 = -1 \pm j$$

or

$$x_3, x_4 = 1 \pm j$$

and by substitution into $f(x) = 0$ it is observed that

$$x_3 = -1 + j$$

and

$$x_4 = -1 - j$$

INDEX